The Book of Facts and Trivia

T0270299

Science

Terri Schlichenmeyer

VISIBLE
INK
PRESS

DETROIT

About the Author

Terri Schlichenmeyer has been reading since she was three years old, and she never goes anywhere without a book. In addition to writing books like this one, Terri is a book reviewer whose works are read in dozens of newspapers and magazines around the country. Her favorite things are nonfiction books (especially trivia books), unusual bookends, and naps. She's considering going pro on the latter. Terri lives on a Wisconsin prairie with one ever-patient man, two "not spoiled" little dogs, and 21,000 books.

The
Book of
Facts and
Trivia

Science

Terri Schlichenmeyer

VISIBLE
INK
PRESS

DETROIT

Also from Visible Ink Press

The Big Book of Facts
by Terri Schlichenmeyer
ISBN: 978-1-57859-720-8

The Book of Facts and Trivia: American History
by Terri Schlichenmeyer
ISBN: 978-1-57859-795-6

The Handy Accounting Answer Book
by Amber Gray, Ph.D.
ISBN: 978-1-57859-675-1

The Handy African American History Answer Book
by Jessie Carnie Smith
ISBN: 978-1-57859-452-8

The Handy American Government Answer Book: How Washington, Politics, and Elections Work
by Gina Misiroglu
ISBN: 978-1-57859-639-3

The Handy American History Answer Book
by David L. Hudson Jr.
ISBN: 978-1-57859-471-9

The Handy Anatomy Answer Book, 2nd edition
by Patricia Barnes-Svarney and Thomas E. Svarney
ISBN: 978-1-57859-542-6

The Handy Answer Book for Kids (and Parents), 2nd edition
by Gina Misiroglu
ISBN: 978-1-57859-219-7

The Handy Armed Forces Answer Book
by Richard Estep
ISBN: 978-1-57859-743-7

The Handy Art History Answer Book
by Madelynn Dickerson
ISBN: 978-1-57859-417-7

The Handy Astronomy Answer Book, 3rd edition
by Charles Liu, Ph.D.
ISBN: 978-1-57859-419-1

The Handy Bible Answer Book
by Jennifer Rebecca Prince
ISBN: 978-1-57859-478-8

The Handy Biology Answer Book, 2nd edition
by Patricia Barnes Svarney and Thomas E. Svarney
ISBN: 978-1-57859-490-0

The Handy Boston Answer Book
by Samuel Willard Crompton
ISBN: 978-1-57859-593-8

The Handy California Answer Book
by Kevin S. Hile
ISBN: 978-1-57859-591-4

The Handy Chemistry Answer Book
by Ian C. Stewart and Justin P. Lamont
ISBN: 978-1-57859-374-3

The Handy Christianity Answer Book
by Steve Werner
ISBN: 978-1-57859-686-7

The Handy Civil War Answer Book
by Samuel Willard Crompton
ISBN: 978-1-57859-476-4

The Handy Communication Answer Book
By Lauren Sergy
ISBN: 978-1-57859-587-7

The Handy Diabetes Answer Book
by Patricia Barnes-Svarney and Thomas E. Svarney
ISBN: 978-1-57859-597-6

The Handy Dinosaur Answer Book, 2nd edition
by Patricia Barnes-Svarney and Thomas E. Svarney
ISBN: 978-1-57859-218-0

The Handy Engineering Answer Book
by DeLean Tolbert Smith, Ph.D.; Aishway Pawar; Nicole P. Pitterson, Ph.D.; and Debra Butler, Ph.D.
ISBN: 978-1-57859-770-3

The Handy English Grammar Answer Book
by Christine A. Hult, Ph.D.
ISBN: 978-1-57859-520-4

The Handy Forensic Science Answer Book: Reading Clues at the Crime Scene, Crime Lab, and in Court
by Patricia Barnes-Svarney and Thomas E. Svarney
ISBN: 978-1-57859-621-8

The Handy Geography Answer Book, 3rd edition
by Paul A. Tucci
ISBN: 978-1-57859-576-1

The Handy Geology Answer Book
by Patricia Barnes-Svarney and Thomas E. Svarney
ISBN: 978-1-57859-156-5

The Handy History Answer Book: From the Stone Age to the Digital Age, 4th edition
by Stephen A. Werner, Ph.D.
ISBN: 978-1-57859-680-5

The Handy Hockey Answer Book
by Stan Fischler
ISBN: 978-1-57859-513-6

The Handy Investing Answer Book
by Paul A. Tucci
ISBN: 978-1-57859-486-3

The Handy Islam Answer Book
by John Renard, Ph.D.
ISBN: 978-1-57859-510-5

The Handy Law Answer Book
by David L. Hudson, Jr., J.D.
ISBN: 978-1-57859-217-3

The Handy Literature Answer Book
By Daniel S. Burt and Deborah G. Felder
ISBN: 978-1-57859-635-5

The Handy Math Answer Book, 2nd edition
by Patricia Barnes-Svarney and Thomas E. Svarney
ISBN: 978-1-57859-373-6

The Handy Military History Answer Book
by Samuel Willard Crompton
ISBN: 978-1-57859-509-9

The Handy Mythology Answer Book
by David A. Leeming, Ph.D.
ISBN: 978-1-57859-475-7

The Handy New York City Answer Book
by Chris Barsanti
ISBN: 978-1-57859-586-0

The Handy Nutrition Answer Book
by Patricia Barnes-Svarney and Thomas E. Svarney
ISBN: 978-1-57859-484-9

The Handy Ocean Answer Book
by Patricia Barnes-Svarney and Thomas E. Svarney
ISBN: 978-1-57859-063-6

The Handy Pennsylvania Answer Book
by Lawrence W. Baker
ISBN: 978-1-57859-610-2

The Handy Personal Finance Answer Book
by Paul A. Tucci
ISBN: 978-1-57859-322-4

The Handy Philosophy Answer Book
by Naomi Zack, Ph.D.
ISBN: 978-1-57859-226-5

The Handy Physics Answer Book, 3rd edition
By Charles Liu, Ph.D.
ISBN: 978-1-57859-695-9

The Handy Presidents Answer Book, 2nd edition
by David L. Hudson
ISB N: 978-1-57859-317-0

The Handy Psychology Answer Book, 2nd edition
by Lisa J. Cohen, Ph.D.
ISBN: 978-1-57859-508-2

The Handy Religion Answer Book, 2nd edition
by John Renard, Ph.D.
ISBN: 978-1-57859-379-8

The Handy Science Answer Book, 5th edition
by The Carnegie Library of Pittsburgh
ISBN: 978-1-57859-691-1

The Handy State-by-State Answer Book: Faces, Places, and Famous Dates for All Fifty States
by Samuel Willard Crompton
ISBN: 978-1-57859-565-5

The Handy Supreme Court Answer Book
by David L Hudson, Jr.
ISBN: 978-1-57859-196-1

The Handy Technology Answer Book
by Naomi E. Balaban and James Bobick
ISBN: 978-1-57859-563-1

The Handy Texas Answer Book
by James L. Haley
ISBN: 978-1-57859-634-8

The Handy Weather Answer Book, 2nd edition
by Kevin S. Hile
ISBN: 978-1-57859-221-0

The Handy Western Philosophy Answer Book: The Ancient Greek Influence on Modern Understanding
by Ed D'Angelo, Ph.D.
ISBN: 978-1-57859-556-3

The Handy Wisconsin Answer Book
by Terri Schlichenmeyer and Mark Meier
ISBN: 978-1-57859-661-4

Please visit us at www.visibleinkpress.com.

The Book of Facts and Trivia: Science

Visible Ink Press®
43311 Joy Rd., #414
Canton, MI 48187-2075

Visible Ink Press is a registered trademark of Visible Ink Press, LLC.

Most Visible Ink Press books are available at special quantity discounts when purchased in bulk by corporations, organizations, or groups. Customized printings, special imprints, messages, and excerpts can be produced to meet your needs. For more information, contact Special Markets Director, Visible Ink Press, www.visibleinkpress.com, or 734-667-3211.

Managing Editor: Kevin S. Hile
Cover Design: John Gouin, Graphikitchen, LLC
Page Design: Cinelli Design
Typesetting: Marco Divita
Proofreaders: Christa Gainor, Suzanne Goraj
Indexer: Shoshana Hurwitz
Cover images: Shutterstock.

ISBNs
Paperback: 978-1-57859-797-0
Hardbound: 978-1-57859-863-2
eBook: 978-1-57859-864-9

Cataloging-in-Publication data is on file at the Library of Congress.

Printed in the United States of America.

10 9 8 7 6 5 4 3 2 1

Dedication

To Mark Moen, my Great American Redheaded Goofball, who saved the day when he said, "You know … there's always 'William.'"

I really do think I'm gonna have to keep you now.

To Darlyne. Biology didn't make us sisters. We did that ourselves.

In loving memory of Ruth Ferguson, Dorothy Molstad, and Carol Keenan. I'm pretty sure I felt you all hovering over my shoulders for these past 16 months, helping me to do a good job. Thank you for making me a better writer and a better friend.

Contents

Photo Sources

Alcor Life Extension Foundation: p. 143.

Brookhaven National Laboratory: p. 87.

Cephas (Wikicommons): p. 24.

John Dedios: p. 168.

Ethan from Manhattan (Wikicommons): p. 134.

Executive Office of the President of the United States: p. 108.

Federal Bureau of Investigation: p. 225.

Franklin Delano Roosevelt Library: p. 102.

Friends of West Norwood Cemetery: p. 206.

Lynn Gilbert: p. 199.

Graevemoore (Wikicommons): p. 221.

Susanne Haerpfer: p. 176.

Toby Hudson: p. 142.

Isaac Newton Institute: p. 79.

David Kessler: p. 158.

Kunsthistorisches Museum Wien (Vienna Art History Museum): p. 30.

Lenapcrd (Wikicommons): p. 188.

Library of Congress: p. 29.

Life Photo Archive: p. 257.

Los Angeles Times: p. 80.

Luna04 (Wikicommons): p. 9.

March of Dimes: p. 100.

Metro-Goldwyn-Mayer: p. 138.

Mme Mim (Wikicommons): p. 178.

Larry D. Moore: p. 42.

Muséum de Toulouse: p. 123 (right).

NASA: pp. 16, 57, 117, 151, 246.

NationalGalleries.org: p. 55.

National Galleries of Scotland: pp. 56, 88.

National Library of France: pp. 136, 217.

National Library of Medicine: p. 194.

National Library of Wales: p. 22.

National Portrait Gallery (London): p. 118.

New York American: p. 92.

New York Public Library Digital Gallery: p. 163 (left).

New York State Historical Association Library: p. 155.

Dylan Parker: p. 64.

Poozeum (Wikicommons): p. 20.

Popular Science Monthly: p. 154.

Public domain: pp. 4, 5, 76, 119, 159, 163 (right), 171, 207, 235, 243, 256, 276.

SAS Scandinavian Airlines: p. 103.

Science History Institute: p. 198.

Shutterstock: pp. 11, 14, 33, 35, 37, 40, 44, 45, 48, 50, 52, 59, 62, 63, 67, 70, 77, 94, 96, 98, 105, 106, 109, 114, 127, 129, 132, 145, 147, 148, 152, 156, 160, 181, 184, 185, 190, 192, 196, 202, 204, 210, 213, 215, 223, 224, 227, 230, 237, 239, 240, 244, 249, 250, 259, 260, 267, 269, 275, 279, 281.

Erik Simonis: p. 82.

Smithsonian Libraries: p. 85.

Henry Spencer: p. 26.

Strickja (Wikicommons): p. 172.

Udenap.org: p. 180.

U.S. Air Force: pp. 263, 265.

U.S. Department of Agriculture: p. 72.

Wapondaponda (Wikicommons): p. 123 (left).

Werner Ustorf: p. 6.

West Virginia Department of Arts, Culture, & History: p. 253.

Andreu Veà: p. 18.

Introduction

Years and years ago, back in the Dark Ages, I had a friend, a next-door neighbor, who was also a boy whom everybody thought was my "boyfriend."

Roll eyes here.

I can't recall how long we were neighbors—a year or three, maybe—and I only have the barest of inklings of what happened to him after his family moved far away, but I do remember this: aside from hanging with me, swapping books and riding bikes, I remember that his sole focus was *science*—specifically, he was all of nine years old, but he wanted nothing more than to become a scientist one day. He devoured serious science books when most kids were reading comic books. We shot off rockets in his back yard, measured stupid stuff in the surrounding fields and trees, captured insects to observe and release, hypothesized about stars and space, and did mind-numbing (to me) experiments. Had there been an internet back then, he never would've left the house. Finally, I literally begged him to stop dragging me into science.

If I knew where he was, I'd apologize. Boy, was I dumb.

I absolutely must admit now—and I hope you do, too, after you've paged through this book—that science is probably one of the most interesting things you could ever know. It's wild, with animals and danger and *possibilities*. It's yummy because you can't have dinner without science. It's funny, if you think about what would happen if something that could "never happen" does. It's sad at the edge of a grave or facing imminent death. Science is *right in front of you,* right next to you—if not directly so, then something you're using is in your grasp because of science. In fact, I'd bet that you're using science in your everyday life, probably more than once or twice a day, and you don't even think about it.

So, I'd like you to think about it. Think about skunks and possums and why you should respect them. Think about why we use someone's feet to measure things and not meters like most of the rest of the world. Think about the odd things your body does and how that makes you both fit in and stand out in a crowd of almost eight billion people. Think about how you'd survive a zombie apocalypse or a dunk in a volcano. Learn about how science can be wrong, how it can be *really* wrong, and how it can be manipulated into GREAT BIG AWKWARDLY WRONG on a national stage. Snag some quick facts that will make you feel like a smarty-pants in minutes.

Couple of things to remember before you get started: Because science changes pretty much daily, some of the things you find in here have already been superseded, proven wrong, proven right, or tossed out altogether because of constantly updated research. If you find something amiss, consider yourself to be a bit of a science genius. You've earned it.

Secondly, remember that science is never boring. *The Book of Facts and Trivia: Science* is proof that even people who were once dumb kids with nerdy neighbors can learn to love it.

—Terri Schlichenmeyer (July 2024)

Book of Facts and Trivia

Science

Firsts and Lasts

Although Ben Franklin (1706–1790) urged his contemporaries to pay attention to certain scientific categories that interested him, rudimentary science courses were first officially offered in U.S. colleges in the early 1800s.

The first woman to be admitted to Massachusetts Institute of Technology (MIT) was Ellen Swallow Richards (1842–1911) in 1870. She went on to graduate with a bachelor of science degree in chemistry. Richards also attended Vassar.

The first human flight happened in 1783 when the Montgolfier brothers, Joseph-Michel (1740–1810) and Jacques-Étienne (1745–1799), put a man in their new invention, the hot-air balloon.

Ada Lovelace (1815–1852) is often said—sometimes controversially—to be the world's first computer programmer. As a mathematician, Lovelace teamed up with Charles Babbage (1791–1871) to invent the analytical engine, which was a fully automatic steam-driven calculator. Lovelace translated and corrected the work of engineer Luigi Federico Menabrea (1809–1896), who had knowledge of Babbage's machine; she then replaced numbers with letters and symbols, thus creating the foundation for a basic computer.

The Code of Hammurabi, which is now housed in the Louvre in France, contains laws on practicing medicine and the penalties if the patient suffers. It was written around 1754 B.C.E.

In the fall of 2022, John McFall (1981–) became the European Space Agency's first astronaut with a physical disability (a "parastronaut"). When he was just 19, McFall lost his right leg in a motorcycle accident.

The first thermometer was invented in 1612 by an Italian scientist charmingly named Santorio Santorio (1561–1636).

Jacqueline Cochran (née Bessie Lee Pittman) (1906–1980), who was the first woman to break the sound barrier in May

Called the "Father of Modern Quantitative Experimentation in Medicine," Santorio Santorio invented several medical devices, including the first air thermoscope to gauge temperatures—i.e., the first thermometer.

1953, learned to fly and achieved her pilot's license in a mere three weeks' time.

The first animals in space were fruit flies sent 68 miles (109 kilometers) above the planet in a U.S. rocket in 1947. They survived their trip.

The first seeds sprouted in lunar soil poked their heads up in mid-2022 in a lab at the University of Florida. Twelve pots with lunar regolith (the dust that coats the Moon) from Apollo 11, 12, and 17 were planted with seeds from a small flowering weed and were placed under grow-lights. All twelve sprouted, though they reportedly didn't thrive as well as did seeds planted in Earth soil.

The first African American to receive the National Medal of Technology was Frederick McKinley Jones (1893–1961), who got his accolade posthumously. Jones worked on several products that enhanced the field of refrigeration; when he died, he owned more than 40 patents for his work.

The first woman to win a Nobel Prize was Marie Curie (1867–1934), who received her prize in physics in 1903. She received a second Nobel Prize, that one for chemistry, in 1911.

The first multiplication tables were used around 2000 B.C.E. by the Babylonians.

Marie Curie was the first woman to win a Nobel Prize, which she did in the field of physics in 1903 and again in 1911 for chemistry.

 In November 2022, the first laboratory-grown blood cells were given to volunteers in a trial. It's hoped that blood cells manufactured from donated stem cells will eliminate any blood shortages that humans might encounter, and thus save countless lives.

Ancient World: You and the Neanderthal?

It may not surprise you to know that our collective Neanderthal ancestors still lurk in our genes: the average human is 2 to 3 percent Neanderthal, and a higher percentage of Denisovan DNA lurks in some Asian peoples. Yeah, so go ahead, make all the jokes you want about it. But the fact is that your friends and family are part caveman—and so are you.

Just what does that mean?

Okay, so let's say you were downtown and you met your Uncle Neanderthal on the sidewalk today. Chances are that if he was cleaned up a little and wearing, say, jeans and a T-shirt, he'd mostly look like any other large-bodied person. He'd be more vigorous, with a wider rib cage and a larger forehead, but he'd be considerably shorter than a modern man; the average Neanderthal man was around 5 feet, 4 inches tall (1.6 meters), give or take an inch or two, and the average Neanderthal woman was about 5 feet (1.5 meters) even. His arms and legs would be a bit shorter, but in clothing you might not notice so much.

Neanderthals were built chunky, sturdy, and thick, unlike the more slender Homo sapiens. While they might have looked fat, they were not at all out of shape but, rather, were very muscular and had a lower center of gravity suited for pulling heavy loads.

Ways Neanderthals Are Better Than You

Let's face it: Neanderthals were big, round guys and that helped them to endure extremes of cold in

FAST FACT

Not much is known about Denisovan people; they were discovered in 2008 but were only officially identified in 2010, and most of what's known comes from DNA and a few small remains. Scientists believe that the Denisovans lived with or near Neanderthals and perhaps mated with them, as well as with more modern *Homo sapiens*. On that note, it's absolutely true that Neanderthals mated with modern *Homo sapiens* too.

northern European and Asian climates; this rotundity might make you think, in fact, that Uncle Neanderthal was a bit on the overweight side. Don't let that cute chubbiness fool you, though: he's great all winter long, but he could adapt easily and quickly to warmer climates, too, if he had to.

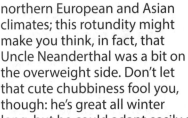

The first Neanderthal bones were discovered in Germany in 1856.

This build, judging by studying where the muscles attached to the bones, indicates that your Uncle Neanderthal was one mighty strong man. Because of those shorter legs and arms we mentioned, and a generally wider pelvis, his center of gravity would have been lower, giving him the strength to haul a small mammoth home for supper at your place, even if it was over a long distance.

Speaking of long distance, studies suggest that Neanderthals were excellent long-distance sprinters, with a lung capacity that enabled them to chase prey until it was exhausted. It was quite a feat of endurance, one that would certainly best most of today's marathoners. Another locomotive benefit your Uncle Neanderthal had was that he could walk a lot longer and farther than you can. Your uncle's eyes were bigger and were better equipped to see in low light. His teeth were often larger than a dentist would see in a modern human, but his sense of smell was rather lacking. This physiology would have meant that he'd have to consume considerably more calories than you do.

Neanderthals, as you may know, had bigger brains than modern humans but not necessarily better gray matter. Intelligence is a tricky word, but suffice it to say that without benefit of compass, how-to book, cell phone, watch, binoculars, and fancy weapons, Uncle Neanderthal and his pals still had smarts. They were able to get around and return to home base, gather random materials to make useful tools for hunting and surviving whenever they needed them, and keep the species alive and thriving for a few thousand years.

> Be a smartypants: Real paleontologists know that the "h" in "Neanderthal" is silent. It's pronounced *nee-AN-der-tall*.

Ways a Neanderthal Is the Same as You

Chances are, if you took your Uncle Neanderthal home for the weekend, you'd find that he was a creative and resourceful guy and may have been gregarious with his people. His lifespan would be comparable to the average *Homo sapiens* of his time, which was somewhere past age 40; that was also the life expectancy of a newborn male Londoner in 1841.

Your uncle would be strong and, like you, he'd have an opposable thumb and hands that could grasp. You'd want to be careful how you shake his hand, though: that forearm gave him extra strength for throwing a spear and chipping away at stone.

Science believes that the ability to laugh became a human trait several million years ago, so your uncle might enjoy a good comedy with you—although you absolutely wouldn't understand one another's languages. He'd have a pretty good memory for certain things that helped him survive and he could probably do very basic math, but don't let him into the kitchen because he doesn't yet have control over fire. Overall, he'd be surprisingly social and very caring—studies show many healed injuries to the average Neanderthal body, which suggests that injury victims were tended and fed while recu-

FAST FACT

Scientists think that Neanderthals and *Homo sapiens* first mated roughly around the Middle East tens of thousands of years ago; the Denisovans may have had a dalliance now and then too. However it happened, Neanderthals left us with certain diseases, including Crohn's disease and lupus.

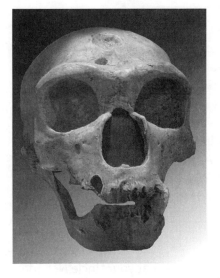

This Neanderthal skull was discovered in La Chapelle-aux-Saints, France, in 1908. Neanderthals had not only a number of physical similarities to modern Homo sapiens, *but also many behavioral ones.*

perating. This gave your Uncle Neanderthal an understanding of family and friends—*and enemies.*

Reproductively speaking, Neanderthals and Sapiens appear to have mated and interbred with abandon on and off for several hundred thousand years, in part because Neanderthals' genes were nearly identical to those of Sapiens, as were their reproductive systems. This is where you get the 1 to 4 percent of Neanderthal genes that reside in your genetics; in fact, and quite ironically given the current population, there's more Neanderthal genetic material walking the Earth today than there was when actual Neanderthals walked the Earth.

Ways That You Are Better Than a Neanderthal

When you consider that your Uncle Neanderthal had to hunt for his own grub and didn't sleep

In 2011, archaeologists discovered an ancient Egyptian pet cemetery that was about 2,000 years old. Dogs and cats were buried there with love, and they also found pet monkeys interred, proving that our not-so-ancient ancestors had beloved, pampered animal companions.

A lot of what we understand about biblical history is thanks to William Foxwell Albright (1891–1971). He is known for the excavation of several biblical sites in Palestine, Jordan, and Arabia, and he furthered our knowledge of the people there.

on designer sheets, he was actually a pretty crafty, resourceful guy—but you're really a whole lot smarter.

Your brain is meant to process information faster and more efficiently than his did. You can grasp the idea, for instance, that the symbols in this book add up to words; he communicated via rudimentary (but still recognizable) drawing. To be fair, you've had the advantage of hundreds of years of invention. Even so, you can fully trust that your Uncle Neanderthal wouldn't have been able to use a computer anytime soon, or learn to drive a car, or use an ATM. You have better cognition overall, and though your uncle was once a part of a small community, you have better socialization and "getting along" skills than he'll ever have.

Physically, well, let's say that Uncle Neanderthal wants to arm-wrestle after dinner. He's a little guy, so this might be awkward, but go ahead and try: His bones are thicker than yours and stronger, and it might take a while, but you could potentially win. Your stamina is better in this kind of physical challenge, and your longer arms give you more leverage.

So does this make Sapiens and Neanderthals different species?

Science has yet to determine that question.

Plants and Animals: Yum, Yum … ICK!

Kudos to the person who first saw a civet cat eating coffee beans and thought, "Wow, what if I roast those beans that come out *the other end* and make something delicious out of them?" That was either incredibly courageous or foolhardy, which is what you could say about the folks who figured out uses for these unsavory, even downright icky things.

- *If you ever* come across a chunk of ambergris while walking along the beach, you might want to throw it back into the ocean and wash your hands—but don't be hasty. It's waxy, black, and quite stinky, but it is a very highly valuable and sought-after material that was used in the spice trade centuries ago and is used in the making of perfumes today. Ambergris is a product that is formed in whales' digestive tracts and regurgitated; in other words, ambergris is whale vomit. Scientists think it's created inside a whale to protect it from the beaks of the creatures it feeds on.

- *Castoreum is another* product used in perfumes and foods such as high-end vanilla flavoring. The natural product consists mostly of secretions from the North American beaver's castor sacs (located near its anus and the base of its tail), but also a little bit of fecal matter, and some beaver urine by accident. It might seem that it would smell horrible but castoreum supposedly actually smells a little like vanilla.

The anal sacs from the North American Beaver produce a chemical called castoreum that has been used as a vanilla substitute. Vanillin (a chemical substitute) is more commonly used these days, however.

Don't sweat its origins, though; castoreum has been approved by the FDA as "safe" for the last 80-plus years. Furthermore, only about 1,000 pounds of castoreum are produced today. Most vanilla that you taste in foods now (if it's not from vanilla beans) is the result of *vanillin*, which is either plant-based (extracted from wood, rice, orchids, or cloves) or petroleum-based guaiacol from crude oil.

The urine from pregnant mares holds relief for menopausal women. The secretion contains estrogens that are isolated for use in hormone replacement therapy.

One more smelly product: musk, which is a secretion from the sexual organ of the male musk deer and of the civet cat, used to be used in perfumes. At the time of this writing, most musk deer are endangered, according to the International Union for Conservation of Nature; much of the musk you'll find in perfumes today is synthetic.

Who doesn't love those sticky-sweet red maraschino cherries? Yummy, right? They're that lovely neon red color due to cochineal beetles, which are crushed to make red dye. This aside, and bearing in mind that eating insects is common and sometimes even good for you, the average American *unwittingly* eats some 2 pounds (not quite 1 kilogram) of insects and maggots each year that have been hiding in their food, according to some estimates.

If you love gelatin desserts, sticky gummy candies, and the like, you should know that you're eating collagen when you enjoy them. Collagen is a product made from the amino acids found in cow, pig, and goat hooves, bones, tendons, and skin boiled down and generally dried into a useful powder.

Wood pulp, or cellulose, is a common filler product used to enhance the use of dried cheese mix, milkshakes, some meat products, breakfast foods, pastries, and other desserts. You'll find cellulose in a surprising number of your favorite fast-food products, which means you eat a lot of trees in your lifetime.

And that civet cat? The Asian palm civet is fed a strong diet of coffee beans that are partially digested and then collected from the cat's fecal matter. The beans are cleaned (thankfully!), dried, and processed to make high-end, high-priced coffee.

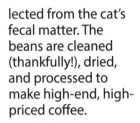

Today's average American eats just under a ton (0.9 metric tons) of food per year. You'll spend several thousand hours eating during your lifetime.

 Legend has it that after his billy goats ate some of the fruits from a certain tree, an Ethiopian shepherd noticed that the animals wanted to do nothing but cavort and play. This, it's said, led to the discovery of Ethiopian coffee.

Physics, Chemistry, and Math: Are You Really Made of Stars?

Scientists have worked hard for hundreds of years to understand what's inside the human body. Of course, you've got a skeleton, blood, and muscles. Dig a little deeper, and you're little more than a mixed skin bag of elements, chemicals, trace metals, bacteria, and genes.

And stars.

To understand how this is true, you first need very basic chemistry and physics lessons.

An element is matter that can't be broken down into something simpler. It is what it is. Each element on the periodic table is given an "atomic number," which helps organize elements by indicating the number of protons in each element's nuclei. This chart is the basis of chemistry, and it's also used in many other areas of study.

Generally speaking, shortly after the universe was born in the Big Bang, the lightest elements—hydrogen, helium, and tiny amounts of lithium—began to slowly form and exist, and after millions of years and the beginning of nuclear fusion, very large stars were created. Inside those stars, a process called nucleosynthesis (a kind of fusion) happened, which created increasingly heavier elements, all the way up to iron; iron is the last element originating in a star. As the largest stars collapsed and exploded into supernovas, they violently expelled the elements they contained to create more complex elements and more (but smaller) stars that made more elements through fusion.

Eventually those elements combined to make an asteroid, which became a planet. The elements dropped to THIS new planet you're on via stardust—an aftermath of a supernova—and began combining willy-nilly to make gases, liquids, solids, and other things that made Earth, Earth.

How? Well, there's still a lot that scientists don't understand about it.

So …

Aside from hydrogen, every element on the periodic chart came from the stars. Considering that everything around you is made of elements, that alone should make you look at the sky tonight with a certain amount of awe.

But then we get to the mind-blowing stuff: You have about a billion billion billion atoms—up

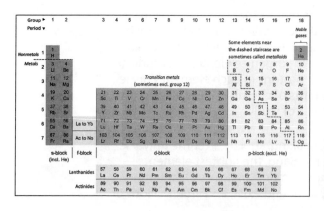

The periodic table of elements lists all the elements known to science, organized by their atomic numbers.

to seven octillion of them—inside your body, give or take a couple. More than 95 percent of them are either hydrogen, nitrogen, oxygen, or carbon. Overall, nearly every atom in your body was once a star; parts of you probably even came through a supernova or two.

Scientists say that over 15,000 pounds (up to 7 metric tons) of stardust falls on the Earth each year. That's a lot of potential people!

If you want to see where you came from, take a visit to the Yerkes Observatory in Williams Bay, Wisconsin. Tours are available, but even though Carl Sagan (1934–1996) and Edwin Hubble (1889–1953) both studied the stars there, the observatory is no longer in use by scientists.

Space Science: Appetizers with Astronauts

When you are roughly 227 nautical miles (420 kilometers) above Earth, it's not like you can just call out for Chinese or pop over to the deli in town. So what do astronauts eat while they're in space?

Tang and Space Food Sticks.

Just kidding, although those were once authentic foods that were really used by American astronauts and also were eventually found on grocers' shelves and in consumers' kitchens.

Although John Glenn (1921–2016) had gone on a fact-finding orbit in 1962, in the mid-1960s there were a lot of unknowns when it came to deciding how to feed a bunch of men (always men then) in zero gravity. Would it be possible to make a nutritious meal for three, something that would allow them to eat it without it floating about the cabin? How could you sate appetites with micropor-

Astronauts (left to right) Charles F. Bolden, Robert L. Gibson, and George P. Nelson prepare food on the space shuttle Columbia.

tions of food that were ultra-lightweight but packed with plenty of vitamins and minerals?

The easy, but likely not very appetizing, solution was to feed the astronauts freeze-dried food in the shape of tubes or cubes, or pureed food in collapsible packaging. The Pillsbury Company was up for that right away, and they came out with Space Food Sticks, which were chewy sticks of sweet goodness, about 4 inches long, pencil-shaped, and packed with vitamins; Scott Carpenter (1925–2013) ate them when he was aboard Aurora 7 in 1962. Tang, a powdered, orange-flavored drink mix, had been invented in 1957 and was taken into orbit by Glenn that same year, so General Foods was on board in *that* category. Other, everyday foods were pureed and eaten as one would eat a tube of toothpaste—nutritious, but not exactly a feast.

FAST FACT

On Earth, a man needs 2,200–3,200 calories to maintain his current weight, depending on his age and activity level; a woman needs 1,600–2,400 per day, with the same factors. In space, calorie needs bump up by as much as a third in order to keep the astronauts' bodies properly nourished for the challenges.

Knowing what we know about living in space today, you can imagine why the early astronauts might have dreamed of a crunchy salad, a juicy steak, or take-out.

As time progressed and technology became better, other industries leaped at the chance to get involved by inventing ways to heat meals, keep them fresh for longer amounts of time, and make them more palatable (even tasty!) while making them more nutritious—safely. NASA and the astronauts benefited from newfangled eating and preparation utensils and a new way of heating water in space.

> We have a lot of reason to thank William A. Mitchell (1911–2004). Not only was he the inventor of Tang, but he was also General Foods' lead inventor for Cool Whip, powdered egg whites, and Pop Rocks candy. On a blech note, he also invented powdered eggs.

Today's astronauts are able to eat almost anything in space that they enjoy on Earth, including mashed potatoes, a crunchy salad, and fresh vegetables that have been successfully grown in space. If it needs to be, food is heated by using MRE (meals ready to eat) technology from the military; freeze-drying is still a common source of meals, and astronauts seem eager and ready to suggest new menu ideas. Meals are made to be appetizing and unique: the astronauts are given a good rotation of meals to avoid menu burn-out.

We really do need to work on that take-out idea, though.

Notable Names: Ray Tomlinson

So how often do you check your email? Nearly half of us check it at least every hour or two when we're not working (more often if we're on the clock), and the average American gets more than 120 emails a day, total. Like it or not, pain in the posterior or higher, email is here to stay.

The late Ray Tomlinson invented the first email system in 1971, using the ARPANET system created by the U.S. Department of Defense.

You have Ray Tomlinson (1941–2016) to thank for that.

Born, raised, and educated in New York, Tomlinson attended Rensselaer Polytechnic Institute, at which he achieved a degree in electrical engineering in 1963. He then attended Massachusetts Institute of Technology for the master's degree he landed in 1965.

Two years later, he joined the tech corporation Bolt, Beranek, and Newman (later renamed BBN Technologies and, since 2009, Raytheon BBN Technologies), where he helped develop computer operating systems and programs. These were the days when computers were huge in size and complicated in operation, and having one in a home was just a dream. Tomlinson was tasked with writing a file transfer program that could do the work of several clerks by transferring files through a system that was already in place. The next step was to change an existing program to send messages to other users of a program on one single computer that users shared. In 1971, Tomlinson took it a step further by adding a code on a platform to allow messages to be sent between multiple computers. By doing this, Tomlinson was the very first person to send an email.

FAST FACT

While it certainly does save time and effort, transferring funds electronically isn't as new as you might think it is. In 1871, Western Union debuted electronic fund transfer. Nearly a century later, in 1974, the Automated Clearing House (ACH) set rules for money transfer. Today, if you get your paycheck direct-deposited, it likely goes through the ACH.

It was unique and unforgettable, a series of letters and numbers at random, sent as a test message. Unbelievably, because he did it without the instruction of his supervisor, neither his employer nor his coworkers thought his new computer program was anything useful or special.

Eventually, though, their minds were changed.

In 2000, not long before home computers became a normal thing, Tomlinson was given the Computer Pioneer Award by the American Computer Museum. That was the first of many accolades Tomlinson received, including one he got posthumously on April 23, 2022, America's first National Email Day.

Ancient World: When You Gotta Go

Let's face it: most of us would—and do—go to great lengths to avoid any kind of fecal matter. Charles Darwin (1809–1882) said that disgust was one of the six most basic human emotions, and when it comes to poop, he got that right—although it should be said that we don't share our propensity to recoil with the entire animal kingdom (dung beetles, we're looking at you).

So why would normal scientists get excited about ancient waste?

Because fossilized poops—known as coprolites—are full of information.

First, the basics about coprolites: identifying them can be tricky. Coprolites are rocks that became rocks when organic matter was replaced by minerals and hardened. So they are rocks, and they look like rocks, but they also kind of look, well, *poopish*, like what you'd expect fresh modern excrement to look like. Because coprolites are no longer organic, they don't smell of skatole, the organic compound that gives poop its smell; instead, they smell like any other rocks you'd find on the ground. If you didn't know what they were, you'd handle them like rocks. Technically speaking, coprolite is a trace fossil with super-high amounts of calcium phosphate, meaning that it's not a substance that kept an animal alive, it's a substance that the animal left behind.

Coprolites are *old*. Dinosaur coprolites can be hundreds of millions of years old—literally as old as dinosaurs. Human coprolites are exciting because they're generally older than most of the discovered objects that an early human might have owned and carried. Human coprolites can also tell things about the general health of an individual and a population, which is likely not possible with a spear, say, or a shard of pottery.

Coprolites like these that formed 30 to 40 million years ago can help paleontologists determine, for example, what animals ate.

The problem with both dino and human coprolites is that sometimes it's impossible to totally, *unquestionably* determine which kind of dinosaur or specific race, gender, or subspecies of human left it, since skeletal fossils and coprolites would very rarely match up in the same spot.

Even so, using technology yields vast amounts of information about the fecal donor.

If there are bone fragments inside the coprolite, it can be assumed that the creature that left the evidence was a carnivore or omnivore. Tooth marks on the excreted bone fragments yield information on dental details and how the bones were eaten. Seeds, pollen, and plant parts indicate a meal that was at least partially made up of vegetation—and it's sometimes possible to see magnified images of those small bits to determine what they were and where they were found before they were eaten.

The size of a creature can be somewhat determined by a coprolite, of course. If you've discovered a basketball-sized poop rock, you can safely assume that it was left by a creature of considerable size and not by a human child.

Even though coprolites are basically rocks, it's sometimes possible to extract DNA from deep inside them. This, along with carbon dating, can track a human and his ancestors, so scientists can learn where he migrated from and what he was hunting and eating along the way. DNA also gives paleontologists a picture of the squatter's health; if there were parasites in the fossilized feces, it's somewhat safe to assume that the person suffered from the same parasites (barring interference from wild animals). On that note, coprolites from ancient dogs are often helpful

> At the time of this writing, the largest human coprolite ever discovered was found in northern England. Believed to have been left by a large man, the rock measures some 8 inches long (20 centimeters) and 2 inches (5 centimeters) wide.

FAST FACT

William Buckland (1784–1856) is considered to be the father of coprology, or the study of coprolites. Originally a theologist (and later, Dean of Westminster), Buckland collected coprolites and wrote the first complete account of a fossilized dinosaur. After analyzing the findings of fossil hunter Mary Anning, he realized what her "bezoar stones" really were. He took credit for coining the word "coprolites," which is derived from Greek *copro* (dung) + *lithos* (stone).

For a while in the mid-1800s, coprolites were mined in England, ground up, and sold for the phosphate inside them. That led to some big bucks for enterprising investors; to this day, there exists a Coprolite Street in Ipswich, England. If you wish to own your very own full-fledged, regular-sized coprolite, it's possible, but beware of the scams. Still, imagine the bragging rights.

when studying nearby human coprolites, since dogs and humans often shared meals.

Plants and Animals: The Birds and the Bees

While it may sound like anthropomorphizing (something scientists hate), animals do have love lives. It's been proven by studies that cows make friends within their herds, and that animals that mate for life definitely grieve if their mate is killed or dies. So, animals are a lot like us in many ways, but they're also quite different.

Male koalas have really attractive pick-up lines, which sound like a cross between bongo drums and a large grunting pig, with maybe a motorcycle engine in the mix somewhere. Scientists believe the *loud* sound, which the koala will make day and night (especially the former), does dual duty, to attract females and to warn other males away.

Fall is generally the time when porcupines in North America mate, and it begins with secretions from the female that include urine and attract male porcupines. Males vying for her attention will fight one another until there's just one left; he will climb up a tree and urinate some more. If he's the one she really hoped would win the fight, she places her tail over her back so she doesn't leave a multitude of quills in his body as they mate.

Other animals that use urine in their mating practices are monkeys, Patagonian maras (an adorable deerlike creature), giraffes, house cats and many big cats, lobsters, male hippopotami, certain kinds of fish, elephants, and male goats.

You'd think scientists might know a lot about the world's largest creature, right? Nope, in reality, mating has never been observed, and therefore, very little is known about the love life or the reproduction process of blue whales or humpback whales.

Bonobos are a notoriously promiscuous species and have been observed having sex with other bonobos of the same

FAST FACT

Porcupine quills don't hurt the porcupine because each quill tip is covered in a coating that contains an antibiotic to protect the porcupine. How often does a stabbing occur? Researchers say clumsy porcupines fall from trees with pretty impressive frequency, so beware of falling objects when walking in the woods.

sex, the opposite sex, and both. It's believed that bonobos use sex as a social glue of sorts, a way to ease tensions, and a method of simply getting along in a hierarchy within a group.

A male octopus has detachable penis, called a hectocotylus, and sperm packets, which he inserts into the mantle of the female octopus during mating. If he's smart, he'll skedaddle then, or she might kill him; if he lives, he'll grow another hectocotylus for another day.

Shortly before he's even a year old, Australia's male antechinus (a small mouse-sized creature) engages in a single day-

FAST FACT

Conservationist William Temple Hornaday's (1854–1937) first job was as a taxidermist at Henry Augustus Ward's (1834–1906) Natural Science Establishment in Rochester, New York. Later, he was appointed chief taxidermist for the National Museum. Hornaday was one of the first people to sound the alarm that the North American bison's numbers were dwindling.

long mating, in which the 3- to 5-ounce (85- to 142-gram) marsupial engages with as many willing females as he can find. For up to 14 hours, the male mates and mates and mates until he's literally physically in tatters and he dies.

Bedbugs are nasty things to find in your sheets, and the male is kind of nasty in the sheets too: When he decides it's time to mate, he climbs on the back of a random female bedbug who's just eaten. He stabs his penis into her abdomen and inseminates her there, because her reproductive organs aren't used for mating but for making eggs *only*. Somehow, the sperm that's injected into the female's abdomen finds her ovaries and fertilizes her there.

While some creatures will pair up with same-sex partners for the raising of offspring (as many birds will do), other animals have been observed mating with same-sex partners. Male bottlenose dolphins will mate with other males, and cows have been known to mount other cows that are in estrus. To throw a monkey wrench into this whole thing, protandry (male to female) and protogyny (female to male) changes are common in fish, depending on the number of partners available at any one time.

Physics, Chemistry, and Math: Wave "Hello!"

There are few things a surfer loves more than a good wave—one that lasts long and rides smooth, with maybe a hint of a challenge. That absolutely can be found near Puerto Malabrigo (also known as Chicama), Peru, on the north coast. It's called a Chicama wave.

Located a little over 350 miles (563 kilometers) from Peru's capital, the Chicama wave (or the *Mamape*, "never-ending wave") is a geological phenomenon first discovered by Chuck Shipman, a surfer from Hawaii who spotted it while flying over Peru on his way home in the mid-1960s. The minute he spotted it, it's said, he couldn't wait to gather a contingent of surfers to travel to what was then a remote area of Peru to check it out.

A Chicama wave starts thousands of miles away, mostly in the Pacific Ocean but sometimes in the Southern (Antarctic) Ocean when weather systems begin to make waves. As waves of similar size and consistency head toward the east side of South America, they tend to congregate together through

Malabrigo is known as the home of the Chicama wave, a special ocean phenomenon that caused the area where it is found to be protected under Peruvian law.

the deep water; because of the shape of the coastline near Puerto Malabrigo and because the water there is still very deep, the waves continue until they are close to the shoreline. Most surfers prefer to surf along any one of four points where the waves separate, which allows for a better ride.

FAST FACT

Without gravity, you would not be able to surf: surfing requires that a displaced mass of water be centered with the force of gravity of the surfer. If you're standing in the middle of a board on the water, not much happens because gravity and buoyancy are perfectly synced, and that's pretty boring. Move physically to the back of the board, and gravity and buoyancy no longer are in sync. The nose of the surfboard goes up, the board twists, and as the surfer adjusts, the waves take over for movement. Voilà!

It's possible for a surfer to catch a wave that takes them on a ride of nearly 2 miles (3 kilometers) of wave. It's rare, but *possible*, to ride a wave (or waves) for up to 4 minutes.

Because of a similar wave that was ruined by construction, the Chicama wave is protected by Peruvian National Law. No one can build anything within one kilometer (0.62 miles) of the bay that will harm the bay itself or affect the wind that makes the Chicama unique.

If you're a surfer, you can find similar waves in South Africa and Australia, but none of them are as long as the Chicama.

Human Life: Ashes to Ashes, Dust to Dust

As the saying goes, there are only two inevitables: death and taxes. But think about it: you don't have to pay taxes if you die.

- *Somewhere between 55* million and 65 million people die in the world each year, depending on the year and the number of disasters.

- *If you die* in New Jersey, you might not be "dead." That's because your family can reject a diagnosis of brain death, based on their religious beliefs.

- *Two chemicals*—putrescine and cadaverine—are what give off the typical stench of death. Both can be bought in chemical form for, among other things, the purpose of training dogs to find the dead in extreme situations.

Future President Ronald Reagan (1911–2004) married his first wife, actress Jane Wyman (1917–2007), in the Wee Kirk o' the Heather chapel inside Forest Lawn Cemetery in New York in 1940.

More than 300 people call Mount Everest their final resting place. Over the years, those people had tried to climb Everest but never made it, mostly due to issues of oxygen and elevation.

The standard depth of burials being 6 feet (2 meters) supposedly dates back to seventeenth-century London. During the Great Plague of 1665, it's said that the Lord Mayor of London decreed the depth so that corpses wouldn't foul the air or sicken anyone further. In many cases today, and because of the ubiquity of burial vaults that protect the corpse from the outside world and vice versa, bodies are buried just 4 feet (just over a meter) below the sod.

The words "cemetery" and "graveyard" are often used to mean the same thing, but there is a subtle difference. A cemetery is a place expressly used to bury the dead. A graveyard has the same basic meaning, but that word is used in conjunction with a church burial ground.

Recent research says that for a few moments after you die, your cognizant brain is aware of what's happening and will continue to work for a short time.

If you add them all up, there are over 140,000 cemeteries in the United States. More to the point, there are over 20,000 *registered* cemeteries and graveyards in the United States— which does not count small family cemeteries, parks that contain graves, or ancient burial grounds. Each year in the United States, more than 1.5 million tons (680,388 kilograms) of concrete are put underground in cemeteries.

Up until recently, more than 4 million gallons (15,000 cubic kilometers) of embalming fluids went into American soil each year. That amount has lessened with the growing popularity of "green" burials and cremations.

One of the ways that can be used to tell if someone is brain dead is to touch the cornea of the eyes with a cottonball. When you're alive, conscious or not, cranial nerves will involuntarily make you blink if an irritation is introduced to the cornea. The reflex obviously doesn't happen when someone's dead. This lack, in conjunction with lack of pulse and lack of detectable motor reflexes, can help a doctor determine brain death.

At the time of this writing, heart disease is the number one cause of death in America. Cancer is the second leading cause of death.

Hart Island, near Manhattan, might be the largest AIDS burial ground in the world. During the AIDS crisis, more than 100,000 victims died in New York City, and many were interred on the Island. Roughly 10 percent of New York's COVID-19 dead are buried there too.

The word "mortician" was coined in 1895 as a sign of respect and to replace the word "undertaker." The word gained popularity when Americans started using the services of morticians rather than preparing their dead themselves.

Human Body: More Death ... Disease

According to the World Health Organization website, the leading cause of death on Earth in 2019 was heart disease, followed in order by stroke, COPD, lung infections, and "neonatal conditions."

At the time of this writing, there are only about half as many board-certified forensic pathologists in America (roughly 500 of them) as there need to be (about a thousand).

President William Howard Taft (1857–1930) was the first U.S. president to be buried at Arlington National Cemetery. He died of heart disease.

It's creepy fun to think of something happening six feet below ground in a graveyard, but it's a total myth that a corpse's hair and fingernails continue to grow after death. What actually happens is the skin on fingers and scalp shrinks, exposing more of the nail and individual hairs.

If a body is buried just a few feet (a couple of meters) below the surface of the Earth without a coffin, it's entirely possible that the person's DNA could remain in the soil for decades, if not longer. If the body is buried in ice, the time frame could be even longer. Much will depend on the elements to which the body is exposed before and after burial.

The famous "Habsburg jaw"—a facial feature the hallmark of which was a long lower jaw and a large lower lip—was probably the result of intermarriage through close relatives within the Spanish Habsburg family in the 15th through early 18th centuries. This genetic flaw was said to be so pronounced in Charles II (1661–1700) that he was unable to chew food or speak clearly.

Eight times more people die on the descent of a major mountain such as Everest than on the ascent. The reason is probably that they've depleted their physical resources getting to the top, so they're slower and thus more susceptible to altitude sickness and the physical maladies from it

Charles II of Spain, one of the many members of the Habsburgs that ruled much of Europe, possessed the distinctive jaw that was inherited by many in his family.

for the descent. In other words, by the time they descend, they're physically not as fit as they were going up.

The very first officially recognized case of Spanish flu was recorded on March 4, 1918, at Fort Riley in Kansas when U.S. Army private and cook Albert Gitchell (1890–1968) got out of bed feeling ill and reported to the camp hospital. The camp doctor recommended that Gitchell quarantine himself, but the bug was loose by then. Within a week more than 500 men had shown up with identical symptoms of high fever and achiness.

The Black Plague, which occurred in the fourteenth century, killed millions of Europeans. Scientists have discovered that those who survived had a set of genetic mutations that helped them fight off the plague's bacteria. Today, those mutant genes that some people got from their ancestors are linked to a higher-than-average risk of Crohn's disease.

It can take a mere 36 hours for a contagious virus to spread globally.

Now for the good news: there are more than 200 different cold viruses, but once you catch an individual virus and recover, your body may have a slight immunity to that particular virus, which might keep you from getting it again.

Generally speaking, the human body can survive a fever temperature of 108°F (42.2°C) for a short time, although there have been reports of people running a body temp slightly higher without dying. High internal temps can cause brain damage, organ damage, and cardiac arrest. As for an outdoor temperature, anything above 109°F (42.22°C) for a sustained period can cause hyperthermia, and the results are the same.

Human Life: Hair Today, Gone Tomorrow

Every day, you brush it, comb it, pick it. You wash it, blow dry it, pluck it, shave it, or basically ignore it altogether. Why do we have hair, anyhow?

Science says that human hair was once a lot like animal hair in texture and color, and it served *Homo sapiens* in the same way that it helped early wolves, woolly mammoths, and saber-tooth tigers: it kept our ancestors warm in the winter, and it kept the sun off our skin in the hot summer months.

But evolution said, "Hey, wait a minute!" In the long run, we didn't need that big, fluffy coat everywhere on our bodies. We evolved better ways of cooling ourselves (namely, sweating), so hair morphed into what it is now: the same number of hairs that we likely had millennia ago, only now thinner and shorter.

Even so, and despite the fact that we no longer run across savannas in search of tonight's dinner, we still need at least some of that fur. Just bear in mind, though, that if someone compliments you on your hair, you might want to ask them to be more specific.

The hair on your head is there to keep you warm in a spot where you tend to lose a lot of heat, and it offers a bit of cushion for bumps on your noggin. Eyebrows exist to keep sweat from dripping into your eyes, and eyelashes shield your eyeballs from most of the dust and debris you encounter without even knowing it. The hair in your nose acts as a first defense in your breathing, so you inhale less debris. Pubic hair and underarm hair are thought to trap scents that are sexually appealing. Arm, chest, and leg hair are also on your body to keep you warm; in fact, each follicle has a tiny muscle attached to it,

FAST FACT

Human hair has been found to be one of our best weapons against oil spills: when it's made into a mat, research shows that it absorbs a surprising amount of oil from the surface of the ocean, helping clean up the mess.

FAST FACT

Hairdressers know all about hair slivers: that's when a short bit of sharp hair, fresh from having been cut, penetrates the outer layer of the skin. Hair slivers (also called hair splinters) are sometimes hard to spot and remove, and they can be every bit as painful as any other kind of sliver.

called the pili muscle, which makes the hair stand up on end when you're cold or scared.

Hair grows everywhere on human skin except on scar tissue, the lips, the palms of the hands, and the soles of the feet—bald individuals notwithstanding.

Here's how it happens: Each individual hair has its very own follicle in your skin, and each follicle contains a hair's root. Fed by tiny blood vessels, the hair grows beneath your skin until it pushes up above the edge of your outer skin layer, where you can finally see it.

By the time they reach this point, the cells that make up each individual hair are dead, which is a little bit creepy when you consider that you're styling

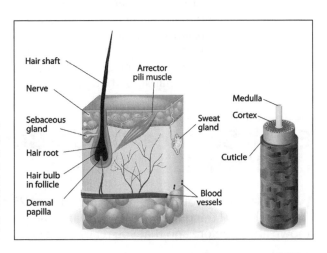

The anatomy of how hair grows from our skin is probably a bit more complex than you might have thought!

Hair shaft
Arrector pili muscle
Nerve
Medulla
Sebaceous gland
Sweat gland
Cortex
Hair root
Cuticle
Hair bulb in follicle
Blood vessels
Dermal papilla

Genetics has a lot to say to Prince Harry: Britain's King William II (1060–1100) was also known as William Rufus—Rufus meaning "red" or "red-headed"—because he had a full head of red hair as a child.

dead cells every morning; on the other hand, dead hair also mean that there's no pain when you go for your eight-week haircut. Most hair, both on your body and on your head, will grow to a finite length before it stops growing, so you can give up that dream of braiding your leg or chest hair now.

Give or take, there are roughly 100,000 to 150,000 individual hairs on your head, and they grow for generally two to six years before they stop growing and fall out. The average person loses 50 to 100 hairs a day from their head, but that's generally no problem: you probably barely notice because that's just a fraction of what you have overall. Happily, another hair starts growing from the same follicle to begin the process over again.

But what if that doesn't happen?

Lots of things can cause a human to become bald. Diet, heredity, hormones, cancer treatments, and medical conditions are but a few reasons. At any age, you can experience bald spots by stressing the follicles with tight hair bands, ponytail holders, or hats. If you ever experience significant, sudden thinning or hair loss, though, it's time to call your doctor.

Earth Science: Ba-da-Bing, Ba-da-BOOM!

Who says what you don't know can't hurt you? There are a lot of geological things beneath the oceans that you might not have known about before and that you really don't want to mess with.

Earth's crust is separated into a number of plates that are either grinding together, sliding next to each other, or separating from each other.

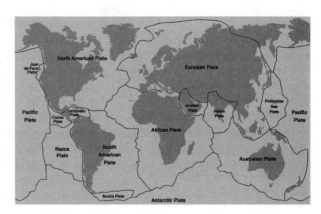

There are around 1,500 potentially active volcanoes on our planet.

More horseshoe-shaped and stretching roughly 25,000 miles (40,234 kilometers) around the Pacific Ocean from New Zealand, northward around Australia and west to Java, then hugging the coast along China including Taiwan and Japan, across the Bering Strait, down the Alaskan and British Columbian coast, and down nearly to the tip of South America is the Ring of Fire. This large area holds some of the planet's most active volcanoes—about 450 of them, most of which are underwater. The Ring of Fire happened because of plate tectonics, especially where two tectonic plates meet. And here you thought it was just a Johnny Cash song.

Speaking of plate tectonics, scientists only recently learned that one of them that runs along the side of the Mariana Trench is covered by hydrous minerals (minerals that hold water) that allow seawater to run deep into the middle of the planet. They knew this was happening—the way the plates are aligned are called subduction zones—but they didn't know how much water is being funneled inward. They still don't know. Suffice it to say: it's a lot.

No, the interior of the planet isn't like an aquarium. The water that goes down the Mariana Trench is coming back to the surface somehow—possibly via the Earth's volcanoes—but a definitive answer hasn't yet been determined.

And speaking of the Mariana Trench, it's located in the Ring of Fire, just off the coast of the Philippines and Taiwan. The

> Another thing you'll find in the spot where tectonic plates are separating: hydrothermal vents, which allow super-extra-heated liquid to escape from the Earth's core.

deepest known trench in the oceans, it's more than 36,000 feet deep (11 kilometers) and 1,500 miles (2,400 kilometers) long. The temperature of the water at the bottom of the Mariana Trench is chilly but not all that bad, considering: it's about 39°F (4°C). You don't want to test that, though: the psi (pounds per square inch) at that level below the surface would literally crush you.

Yes, there seems to be bacterial life that lives at the bottom of the Mariana Trench. What a planet, eh?

Back to the tectonic plates, though: you're on one. That's because tectonic plates are basically just giant slabs of land that sometimes clash underground with other giant slabs of land. That's when earthquakes happen, and if one happens under an ocean, you could see a tsunami, which is a huge wave caused by water displaced by two moving plates. The largest tsunami happened in 1958 in Lituya Bay, Alaska, when a 1,700-foot (520-meter) wave snaked 5 miles (8.5 kilometers) inland and wiped out a lot of trees.

Scientists believe that the giant plates we live on are moving, even now, albeit very slowly. Check back in a few hundred thousand years and you'll see.

Tsunamis are also caused by underwater volcanoes and other big explosions.

Oops, and back to those underwater volcanoes: they are how new islands form. As an underwater volcano erupts, the lava cools and piles up, cools and piles up, until it can be seen above the surface of the ocean. Volcanic islands that are not quite big enough to be seen are called "submarine islands" or "seamounts." The process of a tiny underwater volcano becoming an island you can live on can take millions of years.

In the meantime, there are a number of mountain ranges beneath the oceans, the biggest of which is the mid-ocean

FAST FACT

The Great Pacific Garbage Patch

UNITED STATES

Hawai'i

MEXICO

Garbage concentration in kg per m²

0.1   1   10   100

One unnatural thing you'll find in the ocean is the so-called Great Pacific Garbage Patch, which is actually two swirling vortexes of trash, one off the coast of California and one off the coast of Japan, floating on and in the ocean. Most of what's swirling is plastic, which never biodegrades. Not only is it unsightly, but the trash—which scientists say is almost impossible to measure—causes problems with plankton and algae, which are staple foods for ocean-living creatures. The plastic is also dangerous to birds that ingest it or get entangled in it.

ridge, which runs around the planet for more than 40,000 miles ((64,000 kilometers). Not all of it is underwater—about 90 percent of it is, though.

Hawaii's Mauna Kea volcano is the world's tallest underwater mountain, at roughly 6 miles (10,000 meters) high with a base that lies on the ocean's floor. From bottom to top, Mauna Kea is taller than Mount Everest, the tallest mountain on land.

FAST FACT

In late 1764, diplomat and volcanologist Sir William Hamilton (1730–1803) was sent by the British government as an envoy to Naples. Immediately fascinated by Mount Vesuvius, Hamilton began keeping daily track of the volcano in what would eventually become *eight volumes* of information about the crater, the mountain, and the volcano's activity from 1779 to 1794.

Human Body: Breathe In, Breathe Out

Fun experiment: Cover one nostril and take a deep breath in and out. Then cover the other and do the same. See the difference? It's called the nasal cycle: one side of your nose is on duty while the other side rests, and they switch the workload periodically.

If you're average, you can hold your breath easy enough for 30 to 120 seconds. The longest anyone's officially held their breath (as of the time of this writing) was not quite 25 *minutes*. Trying to best this extreme can be good for you, or it can be bad for you; check with your doctor before throwing a tantrum.

On average, fewer than 1 out of 10 people will sneeze more than four times per day.

Plants and Animals: Biggest to Smallest

The world's largest creature overall is the blue whale, which weighs about as much as 33 elephants and is nearly the length of two large-size school buses front-to-back. Despite their size, the blue whale's main meal is krill, a creature that's generally much less than 10 centimeters long (4 inches).

Growing up to nine feet tall and weighing more than a pro-football linebacker, the common ostrich is the world's largest bird. Be careful how you approach an ostrich: they have a powerful kick, and each foot sports a big middle toe ending in a claw that can disembowel a human in a flash.

The largest insect on Earth today? Depends on who you ask. Most sources say that a number of beetle species are capable of growing to frightening sizes. Thankfully, these insects are not harmful to humans.

Everyone says that whale sharks are docile and sweet, but you wouldn't want to test that—especially once you know that they can grow to lengths that are longer than a pick-up truck and can weigh more than six elephants. Speaking of sharks, eyewitness accounts claim sightings of white sharks that were large enough to bite a large human male in half.

Measuring wingspan, the Atlas moth is the world's largest moth at 10 inches (25 centimeters) from wingtip to wing-tip. The Hercules moth is second, but not by much.

The world's largest rodent is the capybara, weighing in at up to 150 pounds (68 kilograms) and looking somewhat like a cross between a rabbit and a hairy pig. One thing the capybara has in common with a rabbit: he eats his feces, in order to ingest nutrients that his body didn't catch the first time around.

It's a toss-up: Some say that the huntsman spider is the largest arachnid in the world. Other sources say it's the goliath birdeater. Either way, you're looking at a spider that's roughly the size of a dinner plate, give or take an inch or so.

Collectively, the 20 quadrillion ants that populate the planet outweigh humans on the planet.

If you've ever wondered which was the largest animal that ever lived, be prepared to ask almost once a week: that's how often it seems that scientists have been finding fossils

Part of the ta-rantula family, the go-liath birdeater spider is the heaviest of spiders at about 6.2 ounces (175 grams), and it also has the largest body at 5 inches (13 centimeters).

of creatures that supersede ones found before. We have a lot to learn.

On the Other Hand ...

The world's smallest creature overall should be no surprise: it's bacteria, the kind that you can't see without a microscope. (And yes, there *are* bacteria that you can see with the naked eye. *Thiomargarita magnifica*, we're lookin' at you.)

There are several species of dwarf fish that measure themselves in the single-digit millimeters in length. One of them, the *Trimmatom nanus*, measures out at about a single centimeter long. Found in Palau in the western Pacific Ocean, it is considerably shorter than its scientific name.

The smallest mammal in the world is the bumblebee bat—so-named because, at about an inch (2.54 centimeters) in length (sans wings) and weighing about 3 grams (0.11 ounces), the little guy is just a hair larger than a bumblebee.

Smaller than an adult human's fingernail, the *Brookesia nana* is the world's smallest reptile. Two of them, a female and a male, were found in Madagascar in early 2021.

In 2012, Chandra Bahadur Dangi (1939–2015) was confirmed by the Guinness World Record officials to be the world's smallest adult human in recorded history. At 21.51 inches (about half a meter), Dangi, who was born in Nepal,

FAST FACT

In late 2018, a tiny calf was taken to the Mississippi State University College of Veterinary Medicine. The calf was born locally and was in good health, but "Little Bill," as they named him, was small—not quite 10 pounds (4.5 kilograms), or about the size of the average housecat. Alas, the little guy died of an infection at just seven months old.

beat the second-smallest, Gul Mohammad (1957–1997) of India, by not quite a half inch (1.2 centimeters).

The world's smallest feathered friend is the bee hummingbird, which is a little over two inches (5 centimeters) in length, on average. You can find bee hummingbirds in the wild in Cuba.

You'll need a microscope to see the world's smallest insect. He's a parasitic wasp, *Dicopomorpha echmepterygis*, or the fairy fly, the male of which is 0.127 inch long (3.23 millimeters), or about half to a third of a single grain of rice.

Notable Names: Where No Man Has Gone Before

Gene Roddenberry (1921–1991) could have written a TV series about his own life. He was born in Texas and raised in Los Angeles, where his father was a police officer. After a stint in the Army Air Force as a combat pilot, he worked as a commercial pilot before joining his father on the police force. For fun (and maybe for the fame and a little spending money), Roddenberry began writing TV scripts as a freelancer.

He had some successes—a few episodes here and there, a western, a drama, a couple of small credits—before finally launching his own series, *The Lieutenant*, but it only lasted one season. He was not pleased with it anyhow; he called his work until then "very bad shows." In truth, Roddenberry's heart was in science fiction, which, at that point in TV history, was mostly only found in individual television episodes and campy movies.

Nonetheless, he started working on a TV series that he envisioned would be like an outer-space western, and he figured it could "reinvent what an episodic TV show could be." He also wanted to insert a bit of morality, progressive politics, diversity, and then-current events into each episode. With these lofty goals, he approached Lucille Ball (1911–1989), owner of Desilu Studios, with his idea. It's said that she didn't quite understand what Roddenberry had in mind, but she supported the show even before it was a show. Roddenberry set about crafting his new series, but the first episode of his *Star Trek* was awash with problems.

Roddenberry's Spock was all wrong, for one thing; NBC didn't like the first Spock's personality, and the character had badly been miscast with Majel Barrett (1932–2008) in the role; Barrett, who was Roddenberry's mistress then, would later become his wife. In subsequent episodes of the original *Star Trek*, she played Nurse Christine Chapel.

The very first episode didn't have William Shatner as Captain James T. Kirk either; that role was played by

Writer and producer Gene Roddenberry is best remembered as the creator of the Star Trek franchise, which has inspired many young people to pursue the sciences and look to the stars.

Jeffrey Hunter, as Captain Christopher Pike. Pike re-appears in other *Star Trek* spin-offs and episodes.

The show obviously needed a reboot.

This retooling of the show was an accidental bit of good luck. Ball maintained her support, first of all. The revamping gave Roddenberry a chance to hire Leonard Nimoy (1931–2015) to play Spock, a decision that's so entrenched in pop culture that anything else is unimaginable. It gave Roddenberry the opportunity to tweak what he'd done, to add the other characters he'd always imagined in the starship's crew.

In the beginning *Star Trek* had high ratings, but by the end of the season it was almost dead, and NBC, the show's network, moved to cancel it—only to be met with a fair amount of outrage. With Ball's continuing support, fans launched a gigantic letter-writing campaign that delayed the inevitable until the end of the show's third season. All in all, there were 79 episodes filmed.

Fortunately, despite *Star Trek*'s untimely cancellation, Desilu Productions, which was by then known as Paramount Television, had the foresight to license the show for syndication. That bit of fortune-telling is why you can catch original episodes on TV today.

As of this writing, Roddenberry's original idea has spawned a franchise that's worth over $10 billion with multiple television and streaming spin-offs, cartoons, several movies, video games, comic books, fiction books, and countless mugs, statues, toys, tchotchkes, and gewgaws enough to make any *Star Trek* fan happy.

Human Body: Shhhhhhhhhhhhhhhhh

Here's a fact you might not want to hear: we live in a noisy world and it gets noisier every day. Scientists say that unwanted

> **The world's noisiest city is Guangzhou, China. Second-noisiest is New Delhi, India.**

noise—whether it's construction equipment, loud neighbors, traffic, crowds, high-volume concerts, or just regular ambient noise you can't escape—can cause hearing loss and memory loss, makes you feel tense and stressful, could lead to psychiatric issues, and raises your blood pressure. The European Environmental Agency, in fact, estimates that super-loud noise is ultimately responsible for more than 15,000 early deaths per year and contributes to 48,000 new cases of heart disease, via any of the reasons above. In short, noise isn't just an annoyance. It affects your health too.

We humans aren't the only ones.

Noise from increasing ocean traffic is stressful to whales and dolphins, according to marine scientists. Higher volumes cause higher heartbeats in caterpillars, and they cause some birds to lay fewer eggs and to abandon some nests they've started. Other noises cause navigation issues with the creatures that migrate. Fortunately, many (but not all) animals are good adapters.

The five biggest contributors of excessive noise are traffic, both land (up to 80 decibels) and air (up to 85 decibels), nightlife or anywhere people gather so-

Guangzhou, China (pictured), has been measured to have more noise than any other city in the world. The next noisiest were Cairo, Egypt; Paris, France; Beijing, China; and Delhi, India.

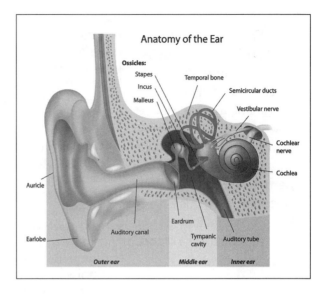

Anatomy of the Ear

Ossicles:
Stapes
Incus
Malleus
Temporal bone
Semicircular ducts
Vestibular nerve
Cochlear nerve
Cochlea
Auricle
Eardrum
Earlobe
Auditory canal
Tympanic cavity
Auditory tube
Outer ear
Middle ear
Inner ear

cially (upwards of 100 decibels or slightly more), construction sites (up to 90 decibels), and domestic animals (your neighbor's dog can bark up to 90 decibels).

Sounds between 60 and 65 decibels, which is about the loudness of a normal conversation, are comfortable for most human ears. Now go back and

FAST FACT

Here's a simple explanation of the magic of how we hear: Sound enters your ear past the pinna, or the outer portion that's designed to work as a cup to amplify the sound waves, which hit the eardrum and make it vibrate. The eardrum is attached to three bones called the ossicles, which amplify the sound even further and move it into the inner ear. There, the sound moves to the cochlea, which contains the organ of Corti. The sound's vibrations in the organ of Corti hits small hair cells that transform the waves into electrical energy that enters your brain and is processed so you don't have to say "WHAT?" again.

Dr. William F. House (1923–2012) invented the cochlear implant. That wasn't his only accomplishment, though: House also solved the mystery of a kind of vertigo that was precluding Alan Shepard from space travel. Because of House's efforts, Shepard was finally able to walk on the Moon.

read the above list again: Anything over 80–85 decibels can affect your hearing and your emotional well-being. Anything over 140 decibels can cause permanent hearing loss.

So what can you do to protect your hearing? The United States tried to help a few decades ago by passing the Noise Control Act of 1972, which authorized the Environmental Protection Agency to address and set limits on common (but growing) outside noises. Alas, the EPA's Office of Noise Abatement and Control was eliminated by Congress in 1981.

Today, OSHA (the Occupational Safety and Health Administration) generally has rules that manufacturers and workplaces must comply with. If you're at home, keep earplugs handy or use your earbuds to tamp down any big sound. Turn down the volume of whatever's too loud, if you can, or leave the site of the noise for a while. Use soundproofing material in your walls. And if all else fails, ask for help from law enforcement, which can advise nearby Noisy Nellies to quiet down some.

Human Life: This Is a Job for ...

Who doesn't want superpowers, huh? It would be fun to jump over buildings or stop speeding vehicles with your hands, for sure—but is such a thing really possible?

 Let's start at the beginning, with America's first major superhero: is it possible that, like Superman, normal humans are really able to be faster than a speeding bullet, stronger than a locomotive, and able to leap tall buildings in a single bound? You gotta love that Man of Steel—but guess what?

Some speeding bullets can go 4,000 feet (1,220 meters) *per second*, give or take; the fastest human could sustain a little over 27 miles an hour (43.5 kilometers), but only for a short time. A railroad locomotive has a carrying capacity of over 12,000 tons (11,000 metric tons); the most a human has ever lifted (at the time of this writing) is a little over a half a ton (907 metric tons). If we say a "tall" building is eight stories, that's about 75 feet (almost 23 meters); the highest confirmed height that a human has ever jumped vertically is 46 inches (1.17 meters). Even with a pole, no one's jumped more than 8 feet, 1/4 inch (2.45 meters) in an official high jump. Can you be like Superman? The Verdict: Nope.

Here's a surprise: Batman doesn't really have any exceptional superpowers to speak of. Instead of any kind of mutation or abilities that came from beyond, he's just a really good crime fighter who uses his intellect and his fighting skills to outwit and catch the bad guys. It just takes a while, is all, but isn't that like real life? Can you be like Batman? The Verdict: Yes, you can be like Batman.

To get an answer on whether you could be like the Incredible Hulk, we need to split his powers. First of all, there are things that could make your skin turn green over time, including wearing copper jewelry, bruising, and a handful of diseases and maladies such as hypochromic anemia and some kinds of organ failure, but these things usually take a while to turn your skin a nice frog color, and they are not instantaneous, not like when the Hulk gets angry and turns that lovely shade. Secondly, you could bulk up by lifting weights, but it generally takes months, if not years, to gain the huge Hulkish musculature you'd need to be like the Hulk. Heavy doses of steroids would bulk you up with muscles faster (though not immediately), but steroids can cause personality changes and anger issues, and it's absolutely nothing you want to mess with, and *those* anger issues wouldn't turn you green. So, can you be like the Hulk? The Verdict: It's not altogether possible and probably not pleasant.

Say you were bitten by a spider that lived in a nuclear reactor. That still wouldn't help you be like Spiderman because the radioactivity might affect the *arachnid* but it wouldn't likely affect the creature's *venom* so it wouldn't transfer web-making powers or any other cool spidery benefits to a

Sorry, Spidey fans, but getting bitten by a radioactive spider will not turn you into a superhero, so please don't go looking for arachnids near your local nuclear power plant.

human—which means having spinnerets on your wrists ain't happening. While it's true that some athletes can *climb* up the side of seemingly smooth monoliths and buildings, the ability to *stick* there like a spider on its web wouldn't naturally be possible. Then there are questions of weight: a spider's web is strong, but unless you're using many, many strands put together, it's probably not strong enough to allow a fully grown man or woman to swing wide over the streets from building to building. Can you be like Spider-man? The Verdict: Not possible.

Right from the beginning, it's obvious that Wonder Woman's just a story: everything about her youth is based in ancient mythology. Could you possess some of her weapons? Well, lassos are pretty common things, but they're not magical. Same with swords, tiaras, and brace-lets—and on that last one, you probably wouldn't be able to deflect bullets with any bracelets you own. And to be faster than Superman? See above. Now, granted ... you could have Wonder Woman's hard-bodied physique, but it wouldn't be the same. Can you be like Wonder Woman? The Verdict: Not the fun parts.

Wolverine is one of those things that make you say, "Hmmmmmm." Originally, he was created as a mutant wol-verine cub from Canada, and his claws came from the gloves he wore. Later writers amended that, making Wolve-rine a mutant human with claws that sprung from within his knuckles. Either way, he's always been an angry guy. So, this is a multi-part answer: No, a mutant cub could not become human. Yes, you can have claws *on your gloves* but because the carpal bones in your hand are small and your wrists are not physiologically set up for them, you would have no room

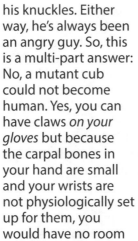

William Kaplan-Altman, also known as Wiccan, appears in the comic series as part of the Young Avengers. Because his superpowers are flying, creating lightning, and casting spells, the verdict is that no, you can't be like Billy.

for them *inside your body*. But yes, you can be grumpy any time you want to be grumpy. So can you be like Wolverine? The Verdict: Not entirely.

There are a surprising number of maybes, if you're looking to be like the Black Panther. Since there are still some things scientists don't know and new discoveries are constantly being made, the idea of there being a hidden culture based on royalty is a distinct, albeit a distant, possibility. Meteors fall from the sky all the time, and beneficial herbs grow in many places around the world. Princess Diana's marriage to the future king of England shows that it's possible for a commoner to become royalty, but it's a slim chance for the average person—especially since royal positions in Wakanda seem to be passed through bloodlines. Also, the whole idea of working with superheroes is kind of iffy. Can you be like the Black Panther? The Verdict: Maybe, but.

Plants and Animals: That's Just Batty!

Ask around and you might find that a good number of your friends suffer from chiroptophobia, to some degree or other—meaning that they have a fear of bats. That's too bad. While bats deserve some degree of caution at times, they're actually pretty interesting creatures.

To begin, when you say "bat," you could be talking about different suborders. Megachiroptera consist solely of the flying foxes (also known as fruit bats) and their Old-World kin, while the Microchiroptera are all the other suborders of bats, of which there are 17 that science has identified, with about 900 species. Megabats are mostly, but not always, larger, with the flying fox's wingspan being the largest at around 5 feet (1.5 meters). Conversely, the Philippine bamboo bat is the smallest bat, weighing in at 0.05 of an ounce (0.85 grams).

Certainly, it's possible to go weeks, months, or even longer without seeing a bat, but unless you live around the poles or in certain oceanic island locales, there *are* bats in your area. They've been a part of the landscape for roughly 50 million years; scientists believe that bats evolved from dinosaurs and may once have had feathers.

While you might think that flying squirrels can fly (they can't; they can only glide), the bat is the only mammal capable of genuine flight. Here's how they do it: If you can imagine this, birds have wings that somewhat resemble a crooked, foldable Z with feathers of different lengths. Bats, conversely, have

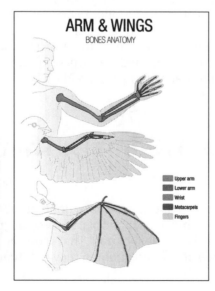

Comparing a human arm with a bird wing and a bat wing gives you a clear picture of how the bones in a bat's wing are actually really long fingers.

ARM & WINGS
BONES ANATOMY

Upper arm
Lower arm
Wrist
Metacarpels
Fingers

wings that resemble arms with very, very long fingers connected by a thin membrane that allows them more maneuverability while in the air. Most birds tend to fly in one direction or another; bats are able to flit and change course quickly in order to catch insects, which are most bats' main food source.

It's a good thing, then, that bats are mostly nocturnal creatures. There aren't a lot of competitors for insect-eating at night, nor are there a lot of bat predators when the Sun goes down. It's cooler at night too, which helps a bat keep its body temperature down; heat is especially detrimental to the bat's wing membrane.

It doesn't matter if it's dark to a hungry, hunting bat: bats are not blind, but their eyesight isn't nearly as important as is their echolocation, which is how a bat finds its prey. Basically, the bat sends out a series of short, sharp sounds that bounce off the insect or prey and return to the bat's sensitive ears, indicating the location of dinner. In addition to this unique way of "seeing" their surroundings, bats use a sort of memory-map of areas they live in or frequent most. It should be noted that some bats catch their meals on the ground or even on the surface of water, some bats are carnivorous, and vampire bats can survive entirely on meals of blood. They are the only mammal to do so. Accept dinner invitations at your own risk.

Bats also use chirps and purrs to communicate to their nest-mates and their young, which are usually born in late spring or early summer, along with most of the other pups in the colony. Depending on the species, there could be one to four pups born while its mother hangs upside down; as soon as a baby bat emerges, Mama Bat catches it and holds on, tucking the infant in a membranous pouch. As soon as the pup is safe inside, it crawls to its mother's nipples and latches on. Within a matter of days, the pup is old enough and strong enough to be left alone while its mother forages, but with some species it takes a village: if a bat mom is out

FAST FACT

Thank your lucky stars that vampire bats don't prefer human blood over animal blood. They'll occasionally sip from a human, but they mostly want blood from other mammals; even so, they won't take much from your cow or horse. Found mostly in Central and South America, a vampire bat feeds by hopping or walking up to a sleeping animal, cutting a small incision in an area of thinner hide, and lapping up the blood that flows from the wound. The danger to the animal isn't in the blood loses, which is relatively negligible. The danger is in what the bat can leave, including rabies.

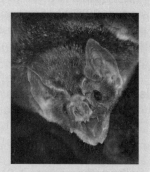

and its pup grows hungry, another female may hear its cries and feed it. Fear not, though: Mom will find her own pup again, out of all the other pups in the nest, by scent.

If all goes well and a bat doesn't sustain an injury or illness, it could live to be 20 years old or older.

While it's true that we need bats to eliminate an overabundance of insects, pollinate plants, spread seeds, and supply fertilizer, despite their size they can be dangerous because of the diseases they can carry. If one gets in your house, open the doors and windows, turn on a light outside, and let the bat find its own way out; or put a thick towel over it and *carefully* carry it outside. Never touch a wild bat otherwise: not only will a bat urinate on itself

Steer clear of a bat-filled cave. A bat will defecate up to 30 times a day, and inhaling bat fecal dust can lead to histoplasmosis, an illness that affects the lungs.

FAST FACT

William Smith Greenfield (1846–1919) was a contemporary of Louis Pasteur (1822–1895), when the latter was working on a cure for rabies. At that same time, Greenfield was working on a vaccine against anthrax.

sometimes, which is icky enough, but bats are the number-one cause of rabies deaths in America.

Science Basics: Measuring the World

- *A billion seconds* is a little more than 31.5 years.

- *The metric system* is the most common way of measuring throughout the world's scientific practices, which makes measurements easy to calculate, no matter what language you use. The metric system is based on multiples of 10.

- *There's a general* argument as to who determined that a foot was 12 inches. It's likely that the first measurement of a foot was literally *a foot,* probably that of a king or other leader. Later indications were that big-footed Romans made the length officially just a hair under 12 inches.

- *In ancient Britain,* barleycorns, or the seed of a barley plant, were used as measurement, with each barleycorn being about a third of an inch. Smart folks chose their barleycorns carefully when measuring.

- *Egyptians used the* cubit, which was the length of the forearm from elbow to middle finger or just a hair under 18 inches. The Royal Cubit was more arbitrary: it was a cubit, with the width of the palm of the ruler at that moment added.

⚛️ *The four most* common measurements that people take today are length, volume, time, and temperature.

⚛️ *To indicate degrees* of measurement in the metric system, scientists use a list of prefixes. It goes like this:

Name	Places	Prefix
Trillion	1 plus 12 zeroes	tera
Billion	1 plus 9 zeroes	giga
Million	1 plus 6 zeroes	mega
Thousand	1 plus 3 zeroes	kilo
Hundred	1 plus two zeroes	hecto
Ten	1 plus one zero	deca
Tenth	0.1	deci
Hundredth	0.01	centi
Thousandth	0.001	milli
Millionth	0.000 001	micro
Billionth	0.000 000 001	nano
Trillionth	0.000 000 000 001	pico

Even more metric measurements:

Quadrillion	1 plus 15 zeroes	peta
Quintillion	1 plus 18 zeroes	exa
Sextillion	1 plus 21 zeroes	zetta
Septillion	1 plus 24 zeroes	yotta
Octillion	1 plus 27 zeroes	ronna
Nonillion	1 plus 30 zeroes	quetta

⚛️ *No doubt you've* heard about horsepower, but the term is somewhat misleading. It was invented in the late 1700s by Scottish engineer James Watt (1736–1819) to compare the amount of work that a steam engine could do to that which a horse could do. A single horsepower, according to the British Imperial System, describes the power it takes to move 33,000 pounds 1 foot in distance in one minute's time (got it?). With that in mind, an average horse works at not quite 15 horsepower.

⚛️ *As if a* "light year," which is nearly 6 trillion miles, isn't far enough, science gives us an even bigger measurement in a

James Watt was a mechanical engineer and chemist who invented the Watt steam engine in 1776 that had a great influence on bringing about the Industrial Revolution. He also devised the concept of the horsepower as a measurement of work potential.

parsec, which, if you're planning a trip soon, is just over three light years.

Beginning in about the eighteenth century, a zolotnik was a unit of measurement used in Russia to measure silver. If you had a zolotnik of silver, you had just a hair over 4 grams, about 0.14 ounces.

Not all countries or cultures measure time in seconds, minutes, and hours. In Thailand a six-hour clock is traditional, and Indian classical music goes by prahars (parts) that range in length according to where the Sun is in relation to the planet. Traditional Chinese timekeeping involved the Sun and the division of the day into small fractions of time, but that also included longer hours.

One of the most unusual units of measure was the smoot. In 1958, the Lambda Chi Alpha fraternity took 5-foot-7 Harvard freshmen Oliver R. Smoot Jr. (1940–) and flipped him end over end to measure how long the Harvard Bridge was in smoots. In case you're dying to know, the bridge is just a tetch over 364 smoots.

There's an easy explanation for why horses are measured in hands: because, back when people did a lot of literal horse trading, they needed a way to measure how high the ani-

The kelvin (K) is the primary unit of temperature used by the International System of Units (SI). The scale is related to Celsius (you can convert Celsius to kelvin by simply subtracting 273.15 degrees). It was named after British physicist William Thomson (1824–1907), who developed the scale and its meas-  urements. His work gained him a title: in 1892 he became Baron Kelvin of Largs, or just Lord Kelvin, in reference to the River Kelvin, which flows through the Baron's home country of Scotland.

mal was at the shoulders (its withers). Hands were, um, *handy*. King Henry VIII set the standard "hand" measurement at the 4 inches that it is now. So, if a horse is 18 hands tall, that's 18 × 4, so you've got 72 inches of animal height, or 6 feet at the horse's shoulders.

According to NASA, a garn is used to measure nausea, named in dubious honor of congressman and space shuttle "citizen passenger" Jake Garn (1932–), who apparently suffered from motion sickness. One garn means that the sufferer is pretty much down for the count.

No doubt you've heard of Six Degrees from Kevin Bacon, which is the number of steps it takes to make a person-by-person connection from an individual to the actor. The same principle goes for the Erdös-Bacon Number, which is the number of steps that connect any mathematician to Paul Erdös (1913–1996), a Hungarian mathematician.

Space Science: To the Moon!

Did you see the Moon last night? Chances are that while it moves in a predictable manner throughout the night sky, it's pretty much in the same place as it was exactly a year ago. But what if it wasn't?

Let's say you could somehow toss a rope around the Moon and move it closer to Earth, just for fun. The Moon's gravitational pull on the tides would be immediate and immense: Low tides would be lower, but the danger is in the opposite—high tides would be considerably higher, resulting in widespread flooding on several coasts. Days would become shorter because our planet would rotate faster; that sped-up rotation and the gravity situation would affect your health because, remember, you are really little more than a bag of water, susceptible to the same pull as the oceans (albeit on a smaller scale). Overall, a closer Moon would be no fun.

What if you pulled too hard and the Moon crashed into Earth all at once—boom? You wouldn't have to worry about finishing this book. You wouldn't have to worry about much at all: chances are that the damage from such a crash would quickly, if not immediately, be catastrophic.

Okay, so what if you blasted as much air as you possibly could, and you moved the Moon away from Earth a few hundred miles? Again, you wouldn't be happy. A farther-away moon would alter the tides, first of all, and would affect our planet's rotation and tilt. It could affect the seasons, although not as much as … (keep reading).

Wow, so what if an asteroid hit the Moon, or some nefarious being decided to blow it up? First of all, the bits and

When we see the Moon it looks quite big, but really it is quite far away from us, and we're in no danger of it crashing into Earth.

250,000 MILES
400,000 KILOMETERS

Frederick William Herschel (1738–1822) was an astronomer who built his own telescope in 1774, and then spent nearly 10 years afterward cataloging the nebulae, constellations, and various clusters of stars. It was Herschel who discovered the planet Uranus in 1781.

pieces of Moon that would fall to Earth would likely cause big problems, depending on where they fell. Even so, you wouldn't want to be around if a half-Moon-sized piece of planet splashed down in the Atlantic or landed in your back yard. Then again, it may not land on Earth at all; it may, instead, send a ring of debris to circulate around the planet, similar to the one that Saturn has, or it might reform as another moon or two. Still, you could count on that debris falling eventually. Earth would stay roughly where it is in relation to the Sun, but with no gravitational pull, tides would be affected and would be diminished. With no moon to dictate it, the 24-hour day-night cycle as we know it would never deviate, but our planet could begin to wobble or even tilt on its side. That could mean big changes in weather along the poles and equator, as well as more ice ages, more frequently.

And what if the Moon started to drift away, like a gigantic outer-space balloon? Trick question: it's already happening. The Moon is moving away from Earth—not far, just a fraction of a foot each year, not enough to notice and therefore not much of a danger at all. But check back in about a million years, and the story may be different.

Human Body: The Bug on Your Arm, Phantom Limbs, and Fetal Kicks

There's something on your neck. You reach back to flick it off, maybe kill it, and—there's nothing there. It happens all the time. So why do we scratch?

The short answer is evolution. Our ancestors needed to develop a way to get parasites and nasty critters off their hides, and scratching was it. In modern times, scratching still does the same thing: It gets rid of a foreign presence that's tickling or annoying you, such as a housefly on your arm or a label on the neck of your shirt. Or that phantom hair.

Secondary answers: It's possible that scratching distracts the nerves that caused the itch in the first place, which makes it feel so darn good. Scratching also protects the skin from something— say, a nasty parasitic fly—that could damage our dermis or make us sick.

So, then, why don't you itch at everything that touches your skin?

Researchers have discovered that inhibitory neurons in your spine work as a sort of middleman between your skin and your brain, letting "scratch now" messages through and minimizing sensations that are not dangerous. So, in the case of an itch, the inhibitory spinal neurons tell the brain that there might be a freeloader on your skin, but just only maybe. If you're merely wearing socks, say, or holding a glass of water, the inhibition in this equation keeps your brain from sending "scratch me" messages to your fingertips because there is no threat or reason to scratch. You should be glad for these neurons; without them, you'd be scratching almost constantly.

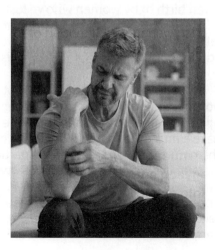

Itches can be caused by real dermatological conditions, but they might also just be your mind playing tricks on you when there is nothing wrong with your skin.

Itches can be caused by many things, including that bug on your arm. It can also come from a psychological issue, from dry skin, from medicines, and from allergies, to name a few things that can trigger you to scratch. Being itchy can be a suggestive thing (which is why you might want to scratch right now) and it can be a contagious thing; if you see someone with an itch, it might make you want to scratch too. Or it can come from a sensation that your brain hallucinates, or from a small breeze that wafts over a single hair on your skin and says to your brain that there's a bug on your neck when there's not.

There are other weird things that your body tries to trick you into believing.

Amputees often feel the sensation that the limb that's been amputated is still there, which is called a phantom limb; phantom pain, therefore, is pain in a limb that's no longer there. The reason for these sensations is not totally known, but it's believed that when an amputee feels a phantom sensation, it's the brain trying to figure out, process, and deal with the loss of what's gone. Phantom sensations, to the relief of many, usually don't last long.

Though they've not been "officially" mentioned in medical literature, studies show that phantom fetal kicks can be felt by new mothers who've just given birth or by women who've lost their babies to miscarriage. In one study, a woman felt phantom fetal kicks even though her "baby" was well into adulthood. Again, you can blame this on the brain.

"Earworms" are what you've got when you've got a song stuck in your head, a phenomenon that up to 98 percent of us suffer from—especially if you have a musical background, tend to obsess about things, are female, or are feeling particularly nostalgic. Not every song becomes an earworm—sometimes you can listen to the radio for *hours* and just one single tune sticks. Other times, all it takes is 10 seconds on "hold" and there you are. For the most part, scientists say, earworms are linked to memory,

FAST FACT

In 2002, Jens Naumann drove a car in the parking lot of the Dobelle Institute in Portugal. That wouldn't be a big deal, except Naumann was blind. He drove thanks to a brain implant invented by biomedical researcher Dr. Bill Dobelle (1941–2004) that simulated sight.

and they cause a loop in the brain; confoundingly, the more you want to get rid of it, the more it seems stuck (called "ironic process"). There are several methods to eliminate an earworm, including replacing it with another song, actually listening to the looped song one time, or just waiting. Most earworms leave on their own.

Just don't be too hasty. Studies show that earworms can be good for you, allowing the mind to wander and relax.

Yep, your brain is a sneaky thing.

You can hear things that are not there, which can be a symptom of a mental illness but is not always. Sometimes, auditory hallucinations are a sign of stress; hypnogogic hallucinations (when you're almost asleep) or hypnopompic hallucinations (when you're coming awake) are so common that up to 70 percent of us have them. The reason isn't totally known, but all of the above originate in the brain.

Studies have proven that if you're shown a picture of something and you hear audio that's similar to what you see, your eyes and your ears will work together to trick you, and you'll "hear" what's in the picture. Seeing different colors can warm you up, cool you down, calm you, anger you, and put you in a holiday mood. Your eyes can also tell you how your food should taste.

And if you believe you can overcome any of these things, welcome to another brain trick: the illusion of control, which overestimates the amount of influence you have on a given situation.

Human Life: What Hollywood Science Fiction Got Right

Catch a few oldies on TV and you might have a good laugh. Seriously, T-Rex and cavewomen never mixed, and your mother's ghost will never inhabit your car. But then again, Hollywood wasn't always wrong when it came to predictions that were scientific.

One of the very first action movies, Georges Méliès' *Le Voyage Dans La Lune* (*A Trip to the Moon*, 1902), is almost cartoonish now, with its literal man-in-the-moon face and a bullet-shaped rocket in its eye. Still, at the turn of the last century, the idea of actually going to the Moon was some really wild stuff. Movies about going to the Moon were pretty common through the entire century, and a lot of them had scary themes. Thankfully, NASA didn't buy into that.

When Samsung invented its first tablet, Apple had objections, but there was proof that the technology wasn't exactly proprietary. In the movie *2001: A Space Odyssey*, director Stanley Kubrick's (1928–1999) characters used what looked an awful lot like a modern computer-based

The charming 1902 movie A Trip to the Moon *was based on the 1865 Jules Verne novel* From the Earth to the Moon.

tablet. Thing is, Kubrick's creation came out several decades before Steve Jobs (1955–2011) presented his gadget: 1968 for the IBM News Pad vs. 2010 for the iPad.

The casual use of computers and computerized features, in fact, were big predictions throughout the latter half of the last century. When *Star Trek: The Next Generation* debuted in 1987, for instance, the characters were using touchscreen computers, while real-life folks were still using 3.5-inch floppy discs, and few had access to the internet.

Airplane II (1982) was a great comedy movie, and one of the gags showed a full-body scanner in an airport, more than 25 years before actual airports had the technology.

In the original Star Trek (1967), characters used devices that looked and acted quite a bit like cell phones—which were invented in 1973 and were not available on the consumer market until about a decade later.

Jim Carrey's (1962–) character in *Cable Guy* really got a lot of things right. The movie was released in 1996, but the character ranted about "integrating" all electronic devices

Robby the Robot, first appearing in MGM's 1956 sci-fi film Forbidden Planet, also appeared in other TV and movie shows in the 1950s and 1960s. At the time, Robby was a groundbreaking design that was much more interesting than the cardboardboxlike robots of earlier sci-fi.

If you are a big fan of cyberpunk movies and books, you can thank Canadian writer William Ford Gibson (1948–), who invented the genre. He's also credited with inventing the term "cyberspace."

in a home, shopping from home, and internet gaming around the world—long before those things were realities.

Ridley Scott's (1937–) Blade Runner featured electronic billboards. The movie came out in 1982; electronic billboards weren't widely installed for another 23 years.

There have been a number of TV shows featuring robots. Robby the Robot was one of the first, as seen on *The Forbidden Planet* (1958); the Daleks appeared on BBC TV's *Doctor Who* franchise in 1963; and B9 was a character on *Lost in Space* in 1965. Can't forget about Data on *Star Trek: The Next Generation* (1987) or the *Transformers* (2002) either. Today's robots are far more advanced than many of those, and we can absolutely expect even more advances in the future.

Of course, who can ever forget Rosie on *The Jetsons* cartoons from 1962 and 1963? Speaking of George, Jane, Elroy, and daughter Judy, the Jetsons introduced kid viewers to video calls (1992, for the ones able to call over regular phone lines), smart watches (the 1990s), flat-screen TVs (1997), drones (early 2000s, for consumer drones), swallowable cameras (2001), and robotic cleaners (introduced to consumers in 2005).

The TV show Small Wonder was about a robotics expert who created a lifelike little girl that the whole family tried to pass off as their "adopted daughter." The astounding thing is that robotics experts have been working on realistic robots for at least 40 years, and artificial intelligence (AI) gets better and better (and a little bit creepier) every year.

Earth Science: Our Weird Planet

So, you think you know all about the planet beneath your feet. If you did, you'd know how unusual and unique it is.

Centuries ago, people believed that the Earth was flat and that if you sailed too far from the continent of Europe, you'd fall into an abyss and never be heard from again. The thing is that they were a little bit right: the Earth *is* flat in some places, but only to your eyes, because of how much you can see of the curve of the sphere. More to the point, the Earth is an ellipse because of the centrifugal force caused by its rotating. Mountains, oceans, and even hills and small valleys have some bearing on the shape of the planet too.

The continents seem to be solidly in place, but the size of them belies their movement. Earth's continents are at the mercy of what happens on the surface of the planet and just beneath it, both of which are constantly in flux because of various influences such as weather, geography, and humans. You can't easily tell, but the continents are moving apart at the rate of a little more than a half-inch per year.

Another thing that's moving is magnetic north. It's moving about 40 miles (64 kilometers) each year, and it's about 600 miles (about 960 kilometers) now from where it was in 1801. Don't get too worried about it, though; the constant movement of the Earth's surface is the cause, and it's common. Fun fact: as far as they can determine, the poles flip every 200,000 to 300,000 years. Just FYI, we're overdue.

Earth's climates vary from hot to cold, wet to dry, and everything in between both. The average temperature overall is 57°F (14°C). Death Valley in

FAST FACT

Born in Scotland but famous for her work in America, Williamina Fleming (1857–1911) was a renowned scientist and astronomer who spent her working life cataloging the stars and other atmospheric phenomena. At a time when women had to fight to stand alongside men in the laboratory, Fleming made numerous discoveries, including the Horsehead Nebula in 1888.

California is the hottest place in North America, but the hottest spot on Earth is in Libya, with air temps that once reached over 136°F (58°C). The coldest place is in Antarctica, where the temp has plunged to colder than −133°F (−92°C). That's a shocking 269°F (nearly 150°C) difference on one small planet.

The driest place on the planet is the Atacama Desert in northern Chile, which receives an average of 1 millimeter of rain a year; it rains in the Atacama maybe two to four times every 100 years. The wettest is Mawsynram in India, which receives just shy of 39 *feet* (almost 12 meters) of precipitation annually. Bring an umbrella, just saying.

Another wet fact: A little over 2 percent of the planet's water is frozen in glaciers. Just over 97 percent of it is ocean and bodies of water connected to the ocean.

It's always raining somewhere on Earth. Stormy fact: Over 8 million lightning strikes hit the planet *per day*.

Gravity is what holds you on Earth, but the gravitational field is not uniform everywhere on the planet. Not that you'll ever go flying off or anything—even if you're standing on the poles, which is where the planet spins the fastest—but uneven terrain on the planet changes its gravity by such a tiny amount that you'd never really notice.

You'd also weigh less at the equator, but not enough to make your doctor happy.

The planet, on the other hand, gets heavier: many tons of cosmic dust fall on Earth each year, landing on the roof of your house, the parking lot at work, and your car every day. Surprisingly, the addition of *you* on Earth had no effect on the planet's weight, because you are an accumulation of atoms that were already present.

Population density across the globe is quite uneven, with the vast majority of people living in Asia, followed by Africa, Europe, and North America.

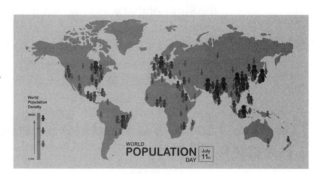

The population of the Earth is not uniformly spread: 6 in 10 Earthlings live on the continent of Asia; fewer than 1 in 10 live in North America. Recent estimates are that nearly 117 billion humans have graced the surface of this planet in the past 192,000 years; roughly 7 percent of all the humans who ever lived are alive now.

If it seems as if you've had a very long day, you're not too wrong. As compared to your great-

FAST FACT

Though there's really no way of knowing for sure, it's estimated that some 117 billion people have walked the Earth since people became people. No one knows for sure how many people our planet can hold today, but population numbers are expected to peak at between 9 and 10.5 billion humans by 2080 before scientists expect that there'll be a sustained population decline.

great grandmother's time, your day *is* longer, since "leap seconds" are added to the Universal Time Clock every few years or so, to keep our clocks and the planet's astronomical time fully synchronized.

Give or take, it takes 24 hours for one spin of the planet each day, and about 365 days to make one complete rotation around the Sun. Our planet is about 93 million miles (nearly 150 million kilometers) from the Sun, not that you'd ever want to get any closer—although that could happen, whether you want it to or not. In a little more than 7 billion years, some researchers say, a collapsing Sun could encroach on the Earth's rotation and vaporize the planet.

Human Body: Thanks, Mom! Thanks, Dad!

If anyone has ever said you've "got your mother's smile," they don't mean it literally, of course. Still, there are a lot of things you can thank Mom and Dad for giving you....

But first, the basics on genetics.

When you were conceived, you inherited two sets of genes, one from your maternal side and one from your paternal side. The vast majority of the genes you have are the same as in all people, but a small percentage of your genetic makeup (about 1 percent) is different. These are alleles, defined as any given gene that everyone has but with your unique, small variations in the sequence of the DNA in the genes. Those differences are what make up your physical appearance and traits you may inherit. Alleles are what make you *you*.

This gene-mixing also serves to lessen the risk of hereditary disease, and it gives future generations a little bit of flexibility in adaptation.

Starting with the top of your head, you can inherit your hair color, eye color, and several facial fea-

FAST FACT

Curiously, a newborn often favors its father more than its mother, no matter which sex the child is. Scientists believe that's an evolutionary tactic to reassure a father that the child was, indeed, his, and to make him stick around for the well-being of the child and its mother at a time when both were most vulnerable.

tures from your parents. All these things may skip a generation; in fact, your skin tone could come from any of the people in your ancestral past.

You might inherit body type from one of your parents, and even some quirks in your physical makeup, but Mom seems to pass on more to her kids than Dad does because she's the one who gives you your mitochondrial genes. The amount of "white fat" and "brown fat" you have affects your ability to gain and lose weight, for example, and that's inherited from your mother. Still, you can overcome that to a degree through diet and exercise.

Your mother's levels of certain hormones leave a mark on you too. If her levels of serotonin are low, you are more apt to develop ADHD. If she has a lower level of *any* hormone in her body, in fact, it could plague you well into adulthood. Mom is the passer of red-green color blindness and hemophilia to her sons, estrogen issues to her daughters, and left-handedness to both. Look at her, and you'll see how you'll age because lines, wrinkles, and such seem to be hereditary. Ask her how she slept, because her sleeping habits are more likely to be yours someday.

The very first person to officially use the word "genetics" to refer to heredity was English biologist William Bateson (1861–1926).

It's not as if Dad didn't give you anything, though.

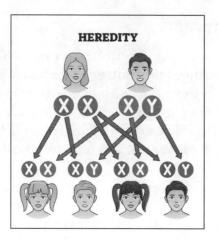

HEREDITY

Your genetic makeup is inherited from your parents in a kind of dice roll that will determine which combination of genes come from which chromosomes that your parents pass along to you.

Your height is somewhat influenced by him, although nutrition also has something to do with your needing extra fabric in your hems. If you have dimples, that's something he passed on to you, and the fullness of your lips will mirror his. Your fingerprint is different from Dad's, but more similar to his than to anyone else's. The structure of your teeth is basically like his, gaps and all. And if there's heart disease in the family, chances are it came down from Dad.

As for things that can go wrong, there are three types of genetic disorders:

Single gene disorder is just what it sounds like: it's when one gene you've inherited is mutated. Examples of diseases caused by a single gene disorder are cystic fibrosis and sickle cell anemia.

Chromosomal disorders are characterized by alterations in all or part of the chromosomes, which are what holds the genes. Down syndrome is a chromosomal disorder.

Complex disorders (sometimes called multifactorial inheritance) happen when two or more genes are mutated. Complex genetic disorders can often be exacerbated by diet, exercise, and general life. Heart disease, diabetes, and some cancers are examples.

Overall, when it comes to genes, a lot can go right and some things can go wrong. You just have to keep your sense of optimism. Yep, that and pessimism are inherited traits.

Plants and Animals: Meet the World's Largest Organisms

Time to blow your mind: if you want to see one of the world's biggest plants, you need an airplane ticket and a snorkel.

Hiding in plain sight for more than 4,500 years, a bed of seagrass is currently considered to be the world's largest living plant. The greenery, believed to be a cross between Poseidon's ribbon weed and some as-yet-undetermined species of seagrass, is located right off the coast of Australia in Shark Bay and was only discovered about a decade ago. Scientists say that just *one plant*—a single seed— went clonal (meaning that it reproduced without sexual fertilization) and is responsible for covering more than 100 miles (160 kilometers), a feat that was proven when researchers took genetic samples from several spots along the sea floor over a period of seven years and found almost no genetic variations.

More mind-blowing: two more big-organism contenders can be found in the United States. In a way, they're also the oldest living things in the coun-

> **FAST FACT**
>
> There are more than 70 different kinds of seagrass in the world's oceans, and while few people ever consider a plant that grows on the bottom of the ocean, you should. Seagrass is a source of food for many undersea creatures; it provides cover for juvenile marine animals; and it helps clean the ocean floor.

try. And they're still alive *and* you can still visit them. But don't wait too long.

In 1968, scientists Burton Barnes (1930–2014) and Jerry Kemperman discovered something that surprised them: tucked away in Fishlake National Forest in Utah was a stand of clonal quaking aspen trees that had been quietly growing for maybe the last 80,000 years, possibly considerably longer. Aside from its incredible age, though, there's more: genetic testing proved that the entire thing, every single tree and every leaf, was actually a part of one individual male organism, connected by a huge and very heavy root system. They named the tree "Pando," which is Latin for "I spread."

Just how big is Pando? At its largest, it weighed some 13 million pounds (nearly 6 million kilograms)

Despite appearances, this isn't a grove of individual trees but one gigantic quaking aspen tree sharing a root system. It is estimated to be 80,000 years old.

FAST FACT

More argument: a gigantic fungus in Oregon was discovered in 1998 and many say that *that's* the largest organism. The "Humongous Fungus" lives on over 2,300 acres (9 square kilometers), which is roughly 1,665 football fields in size. It's believed to be as much as 8,600 years old.

and took over almost 107 acres (nearly half a square kilometer); originally, more than 40,000 trees were included in the organism's vast spread, and some of them first sprouted when Grover Cleveland was alive.

It's believed that that massive root system is what kept Pando around so long: whenever there was a draught or a fire, lightning strike, or other natural disaster that might have destroyed any other tree nearby, Pando took any loss it might have sustained and just sent forth another sapling. No big deal, right?

If you want to see a gorgeous stand of quaking aspens with their fall colors on display, head some September or October for Williams, Arizona, about an hour or so from the South Rim of the Grand Canyon.

It is now. Then-and-now photos show that large areas of Pando have died off and the organism is struggling. The blame has often been laid at the feet of native wildlife, primarily elk and mule deer that dine on young aspen. It doesn't help that it's been awfully dry in the geographical area where Pando spreads; also factored in is the age of many of the trees, which means that there are fewer new saplings sent forth. And then there's the human factor: visitors have loved Pando literally almost to death, and ranchers have been allowed to graze livestock in the area, which may mean more stress for the organism.

And so researchers and biologists are working to save Pando, but who knows? Meanwhile, if you want to stand on the edge of a massive natural organism, don't wait too long....

Human Life: How to Get a Kid Interested in Science

Naturally, if you're having fun reading this book, you'll want your favorite kid to enjoy the same kinds of things you like. So, what do the experts say about keeping a kid interested in STEM (science, technology, electronics, math) projects?

First of all, make science fun! The words that make up the acronym STEM can have a negative connotation for some kids, even though the subjects themselves are super interesting. Present science as a fun thing, more like play.

Go outside and explore. Show your child that there's a whole world out there, even if you're just sitting on one small patch of grass. Look for unique bugs. Pick up sticks and leaves. Watch it rain and learn together where rainwater goes. Teach your kid to garden, to be environmentally conscious, to bird-watch, to look for interesting natural phenomena such as rock outcroppings or unique flora. Take a walk around the block and make it a game to see something unusual and naturally occurring.

Let your child interact with animals, but do it safely. Take them to the vet with Fluffy or Fido and encourage questions to a patient veterinarian. See if a local animal shelter has any volunteer opportunities geared toward kids. Feed squirrels or birds in your backyard. Find a friendly zookeeper, pet owner, or farmer who has patience enough to tell a kid anything they want to know.

Let them ask "What?" questions, as well as "Why?"

When your child notices things and doesn't understand them, take the time to do experiments. Make a mess. See what happens if you make a different kind of mess to get the same results.

Furnish your child with toys that invite STEM-like imagination. Blocks are great for little kids; so are building toys and simple kits to make cool gadgets. Play games that encourage money-counting, mathletics, learning the parts of the skeleton, or any other fun subject you can think of to play while you're waiting and not otherwise occupied. Little kids love dinosaurs; tap into that love and see what happens.

Bake or cook together. Not only will you have a treat afterward, but you'll teach fractions and measuring too.

Don't ever "dumb down" a word. You'll both be glad if you use "real" words for real things.

Take your kid to the museum. Watch nature shows on TV that are age appropriate.

Help your child to learn by looking things up and making learning an open-ended thing.

Plants and Animals: Menagerie, Part 1

- *Butterfly wings consist* of super-thin layers of protein called chitin, which is attached to a system of veins to nourish the chitin. Atop this are layer upon layer of scales, similar to the shingles of a house, that give the butterfly its color.

- *The world's largest* pitcher plant, *Nepenthes rajah*, which is found in Borneo, consumes meat. Scientists have even seen evidence that it will trap rats for dinner.

- *A platypus doesn't* have a stomach; its gullet is connected directly to its intestines. Same with spiny echidnas and lungfish.

- *The Cymothoa exigua* is a real louse. When the female enters the gills of a fish, it severs the blood vessels in the fish's tongue and waits until the tongue falls off. After it does, the louse attaches itself to the stub of what's left and becomes the fish's tongue. It doesn't appear to hurt the fish much, although it's said that some fish tend to become a little underweight.

The **Nepenthes rajah** *is an endangered pitcher plant native to Borneo and is the largest plant of its kind.*

Scientists are still arguing about the existence of rat kings, a phenomenon in which a group of rats becomes so clustered that their tails knot together. Some say that rat kings don't exist; others say that they're very rare but very real.

Up to a billion birds are killed in the United States annually because they collide with windows.

There is no difference between a beak and a bill. They both describe the thing that holds a bird's nostrils and mouth—although ornithologists tend to use the word "bill" more than "beak."

Hippos are considered to be one of the most dangerous land animals alive. Literally, a hippo will attack with no warning, and if you're small enough they can swallow you whole. That's gonna hurt.

Kittens, piglets, and rabbit kits often show favoritism for a certain nipple when nursing and will predominantly head for that particular spot when Mom comes around.

Forget badgers, wolves, sharks, and anacondas. Scientists think that the deadliest (but not necessarily the most dangerous) thing in the universe is a black hole.

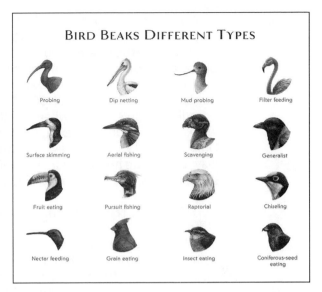

BIRD BEAKS DIFFERENT TYPES

Although there is technically no difference between a beak or a bill on a bird, ornithologists tend to use the word "bill" when talking about flatter, fleshier beaks like those on a duck versus pointier beaks like those on a hawk.

 Moray eels generally only grow to around 5 feet (1.5 meters) in length, but at least one species can more than double that. Morays are not poisonous, but their bite can leave you with a serious injury, and it's best not to hang around them. So: when long fish bites your arm and it causes you harm, that's a moray.

Physics, Chemistry, and Math: Rules of Life and Lab

We need laws to keep the peace, maintain civility, and get by in normal society. Here are a few "laws" that scientists use—and for a bonus, a few fun ones that economists use.

It's not just a television show; the Big Bang theory says the universe began with one great big, very large, colossal explosion. Edward Hubble (1889–1953) and Albert Einstein (1879–1955), among other big-name scientists, believed this to be so.

If you think there must be life on other planets, then you're using the Fermi paradox and the Drake equation, which work hand-in-hand. The former asks why we haven't been visited by extraterrestrial life yet, and the latter attempts to estimate the number of actual existing alien communities that could come for a weekend or two. These scientific "laws" came from physicist Enrico Fermi (1901–1954), who supposedly wondered aloud where the little green men were, if they did, indeed, exist; and from Dr. Frank Drake (1930–1922), who devised a mathematical way to estimate how overrun we could become should aliens visit. The equation looks like this:

$$N = R_* \cdot f_p \cdot n_e \cdot f_l \cdot f_i \cdot f_c \cdot L$$

Where N = the number of civilizations in the Milky Way galaxy with which communication might be possible; R_* = the average rate of star formation in our Galaxy; f_p = the fraction of those stars that have planets; n_e = the average number of planets that can potentially support life per star that has planets; f_l = the fraction of planets that could support life that actually develop life at some point; f_i = the fraction of planets with life that go on to develop intelligent life (civilizations); f_c = the fraction of civilizations that develop a technology that releases detectable signs of their existence into space; and L = the length of time for which such civilizations release detectable signals into space.

FAST FACT

A Helmholtz resonator, named after German physicist Hermann von Helmholtz (1821–1894), is an instrument that measures musical tones. You can make your own simple Helmholtz resonator by blowing gently across the top of a bottle; the noise that occurs when the air inside the bottle resonates will vary according to the size of the bottle's mouth, the speed of your blowing, and the amount of air inside the bottle.

Kepler's laws of planetary motion give you a three-in-one method to describe the way the planets orbit the Sun. Created by Johannes Kepler (1571–1630), it states that planets orbit the Sun in an elliptical fashion with the Sun as their centerpiece, that a line between a planet and the Sun is equal over an equal amount of time, and that a planet's orbital period and its distance from the Sun are related.

Sir Isaac Newton (1643–1747) gave us his laws of motion that state, first, that an object at rest remains at rest, but when the object is set in motion it will keep moving unless acted upon by an outside force. The second law states that the time frame for the object to stop moving is equal to the magnitude and force of what stops it. And third, for every action, there is an equal and opposite reaction.

Do you start to get hungry when you see a clock approaching noon? Congratulations, you're a victim of conditional learning. The most famous example of this is Pavlov's dog, who was taught to associate the sound of a dinner bell with its actual meal. Pretty soon, the dog salivated over the *sound*, and not the *scent* of dinner, an example of classic conditioning.

Sir Isaac Newton established the basis for what we call classical mechanics, cofounded the field of calculus, and made huge contributions to optics, among many other accomplishments.

Occam's razor, in a few words, states that the simpler the answer or experiment, the better. It's also known as the principle of parsimony. "If you hear hoof beats, look for horses, not zebras" is a classic way of understanding this "law" in action.

Eureka! That's supposedly what Archimedes (287 B.C.E.–212 B.C.E.) shouted as he understood his buoyancy principle that explains how an object that is submerged or partly submerged in water will replace water of an equal weight.

Here's one that seems especially apt these days: the Dunning-Kruger effect happens when a person's deficiency in knowledge of a certain subject causes them to severely overestimate their competence in that subject. We can thank psychologists David Dunning and Justin Kruger for this description.

The Streisand effect explains the phenomenon in which the attempt to conceal or divert attention

FAST FACT

Will Keith Kellogg (1860–1951) had a theory that good health could be created by eating only healthy foods, heavy on vegetables. In his younger days, Will worked as a bookkeeper for his older brother, John, who ran the Battle Creek Sanitarium, while Will developed his theory and created a way to bake the Sanitarium's breakfast cereal grains into flakes. In 1906, Will left his brother's business and went into the cereal business by himself. You can still see his name at breakfast every morning, right next to your bowl.

from something backfires spectacularly, resulting in people noticing the item even more than they would have otherwise. This "law" was named for singer and actress Barbra Streisand's attempt in 2003 to prevent the publication of a photo of her home.

Plants and Animals: Smallest to Biggest

Here's a fact for you: we cannot live without plants on this planet. From the largest to the smallest to the most popular, here are a few facts about the green things in our world.

- *The largest trees* in the world are the giant sequoias, which are native to California and the western part of North America. It's pretty common for a sequoia to grow to a height approaching 300 feet tall (91 meters).

- *Measuring over 3* feet (1 meter) in diameter and weighing more than 10 pounds (4.5 kilograms), the *Rafflesia arnoldii* is the world's biggest flower. You don't want to give one to your sweetie, though, and you definitely don't want to try to grow it in your garden: this gigantic flowering plant is also called the corpse plant because the flower smells like something died in it.

- *The most common* kind of grass you'll find in your lawn will likely depend on where you live: some varieties do better in warmer climates; others do better in cooler or wetter or drier places. According to Scotts.com, there are about a dozen different grasses used on lawns in the United States, and most lawns contain a mixture of two or more of them. For the record, American cemeteries also seem to rely on a variety of grasses for cover.

- *So, you say* you love to jump in a pile of leaves after your annual fall raking? Check this out: the *Raphia regalis*, a type of palm tree that grows in some places in Africa, sports individual leaves that may grow to more than 80 feet long (24 meters) and up to 10 feet (3 meters) wide. Yeah, rake that.

The world's smallest tree is the dwarf willow, which tops out at a little over 2 inches (5 centimeters) in height when fully mature. If you've missed it, you're not alone.

The tiniest flower in the world is the watermeal, which measures roughly the size of the head of a straight pin. It's thought that if you scooped them up from their watery home, you'd need around 5,000 of them to fill a thimble.

While enormous pumpkins can be cultivated and raised with proper care, tropical yams from the genus *Dioscorea* will regularly grow to be 60 pounds (27.2 kilograms) or more, although they're usually harvested long before they reach that measurement.

Ancient Sumerians and Babylonians kept plants in their homes for ornamental purposes some 2,500 years ago.

Zones for planting are helpful guides to assist in ensuring that the plants you want to grow are hardy enough to withstand climate extremes in your area. Each zone differs by 10°F (−12°C) from the zone above or below it, so look carefully before you pick your seed or sapling.

America's oldest tree is thought to have been Prometheus, a Great Basin bristlecone pine that lived in Wheeler Peak, Nevada. Before it was cut down in 1964, Prometheus was

FAST FACT

Alberta Maria Wilhelmina Mennega (1912–2009) was a Dutch botanist best known for her studies with woody plants. Originally trained to be a physiologist, Mennega switched her studies to wood because there was a need for that kind of teacher at the college at which she taught.

believed to be nearly 5,000 years old. The oldest living individual (not a colony) tree in America today is in eastern California—Methuselah, also a bristlecone pine tree, thought to be more than 4,800 years old.

The oldest potted plant is believed to be an Eastern Cape giant cycad that lives in the Kew Gardens in the U.K. Brought back from South Africa in 1775 by botanist Francis Masson (1741–1805), the plant has now achieved a height of some 12 feet (4 meters) and weighs a little over a ton.

Orchids are among the most expensive plants in the world, with prices reaching six figures for some of the rarest kinds. Rare roses, ancient bonsai trees, and the Queen of the Night succulent from Sri Lanka will also set you back a few weeks' pay.

Human Body: What Your Gut Says

So your day started out all wrong and you wonder if you have a brain in your head. You do, and another one in your gut, according to science.

The first thing to know is that you are not alone. Your body is home to trillions of microbial organisms on the outside and on the inside, and the mixture you carry around with you is unique to you alone. Part of it came through DNA from your parents when you were conceived, and some of it was given to you through birth; your environment at various points in your life gave you some of the bacteria, as did your diet and your habits.

The highest number of those organisms, known as your microbiome, is found in your colon. What you have there may be beneficial bacteria, or not; viruses and bad bacteria lurk there too, but the good guys and the bad guys coexist pretty peacefully most of the time. If something changes the bal-

> **About 30 percent of your bowel movement is dead bacteria from your digestive system. The rest is undigestible food, cholesterol, some minerals, and fats.**

ance, however—something like a newcomer in your personal bacteria world, an extended use of antibiotics, illness, or a radical change in diet—it could cause you to get sick.

It hasn't been long since scientists discovered this connection.

Though scientists call it a "brain," what's in your gut is not like the brain in your head. The enteric nervous system, sometimes known as your "second brain," consists of many millions of nerve cells that line your body from your throat to your rectum, and it controls digestion. From the time you swallow until you head to the potty, your ENS is at work.

But that's not its only job: it also communicates with the bigger brain in your head, 24/7, via neurotransmitters. This is called the gut-brain axis.

The simplest explanation is that your brain and your gut communicate both ways, via your enteric nervous system, but also involving other systems and parts of your anatomy. Among other things, this communication helps you to know when it's time to eat and when it's time to go to the bathroom. You may see this connection in action, say, if you're stressed and it gives you diarrhea, if you suffer from depression, or if you suffer from digestive diseases such as irritable bowel syndrome.

> **Studies suggest that when immigrants arrive in a new country, their microbiome starts to adapt to the new country (and new diet and new environment) nearly immediately.**

Even though science knows all this about how our exquisite bodies work, the bottom line (no pun intended) is that there's still a lot to learn about the second brain in our guts, and studies are being done almost constantly. Stay tuned!

Physics, Chemistry, and Math: Serve It Up on a Dish of ... What?

Your grandma is so proud of her collection of glassware. She has it in a special cabinet with special lighting that makes it glow. So, what gives?

Chances are that she has a fine collection of glassware with uranium in it.

Uranium—like, the stuff that you find in nuclear reactors? That can't be safe, can it? Nope, it's not safe. The main danger of exposure to uranium is the potential for kidney damage.

Although the very first use of uranium in glassware is unknown, the metal was officially discovered in 1789 by German chemist Martin Heinrich Klaproth (1743–1817) in a batch of pitchblende. Uranium was so named because the planet Uranus had recently been discovered, and one name led to the other. Scientists didn't know then that uranium is a

German chemist Martin Klaproth discovered uranium in 1789.

FAST FACT

To get the full effect of Grandma's uranium glass-ware, turn a black light on it. The black light gives off UV rays that are absorbed by the glassware and re-emitted at a different wavelength. Check it out: the glassware will begin to glow!

relatively common heavy metal found on land and in the oceans.

So anyhow, Klaproth discovered it. And it didn't take long for glassmakers to start adding uranium to their wares because it glowed, and how totally cool is that? Uranium was used in jewelry such as neck-laces and earrings, but for sure, uranium glassware was the most popular item made; from about the 1830s to the mid-twentieth century, it was produced all over Europe and in the United States and Canada.

By the 1940s, governments worldwide began cracking down on the use of uranium on frivolous items such as dinnerware because it was needed for more important things like weapons. Production of uranium glassware was all but halted during World War II and the Cold War, but late in the last century some manufacturers picked up production. Even so, uranium glassware sales were never the same as they were before World War II.

The very good news is that the amount of ura-nium in antique and collectible glassware and dishes is tiny, almost ridiculously small, and it's thought to be safe to handle. Chances are, though, that uranium glass *might* set off a Geiger counter. The Environmental Protection Agency recommends that you not use certain types of uranium glassware (specifically orange or red Fiestaware or yellowish Vaseline glass) with anything you'd want to eat or drink—and be sure to dispose of uranium glass properly if it breaks or chips.

FAST FACT

One of the first nuclear physicists to work on the nuclear bomb was American Willy Higinbotham (1910–1994). Not only did he help create the bomb, but in 1958 he created *Tennis for Two*, which is thought to be one of the world's first video games.

Notable Names: Mary, Mary, Quite Contrary, Part 1

So far, as you've undoubtedly already noticed, William, Willie, and Wilhelmina are scattered all over *The Book of Facts and Trivia: Science*. But it was almost all about Mary.

If you were an adult woman living in the very early 1800s, your options for a future were slim: most women were discouraged then from doing the jobs that men claimed. And then there was Mary Somerville.

She was born in late 1780, the second of four children to survive in a somewhat aristocratic family in Scotland. The Fairfax family was considered to be upper class, which doesn't mean much; they were not rich by any means, but Mary's brothers were both given ample educations. It was thought to be a waste of time to extend the same to Mary and her sister because they were girls. Still, Mary was taught to read, but not to write.

When she reached age 10, Mary's parents apparently had a change of heart and sent Mary away to a boarding school. She decided on her own, once she came home, to continue her education by herself, and she devoured the family library. Her uncle encouraged her and taught her Latin.

By 1804, Mary was married to a cousin, and though he didn't discourage her, he didn't entirely approve of her love of and desire to study mathematics. After he died, Mary took advantage of her time, devoting it to her beloved mathematics.

Her second husband, William Somerville (1771–1860), another cousin, was proud of Mary's work, and that gave her room to add geology and botany to her studies.

The Somervilles moved to London in 1816 and began to socialize with the city's more eminent

Mathematician and astronomer Mary Somerville was considered one of the most brilliant scientists of her day.

scientists, astronomers, physicists, and mathematicians—and Mary's already impressive education expanded tenfold. In 1827, she was asked to work on condensing a five-volume interpretation of the mechanics of the solar system so it could be comprehended by and sold to average Londoners.

In 1834, she did it with a book on the physical sciences; her third book was a textbook on geography; her fourth book was on molecular science. She was elected to the Royal Astronomical Society in 1835, to the American Geographical and Statistical Society in 1857, and to the Italian Geographical Society in 1870, and she began lending her support to women's suffrage at around this time.

And she never stopped writing. Her last book, an autobiography, was published in 1874, two years after her death.

Decades ago, before medicine was modern, if an epidemic hit an area hard, it often took a lot of lives, sometimes so quickly that doctors had little time to discover the details before the disease moved on or away.

And so it was with Mary Mallon, whose story starts in Bristol, England, in 1873, four years after she was born. That's when Dr. William Budd (1811–1880) published a paper demonstrating that typhoid, an intestinal disease, was all too easily transmitted through drinking water in an unusual manner.

Say someone had dysentery or a similar malady and they drank water from a container that was used by multiple people. The saliva of the sick person could contaminate the water when the ladle or cup was re-dipped for the next thirsty user. A nastier method of

In 1836, Mary Somerville wrote that she had trouble calculating the exact position of Uranus, and that there might be another planet behind it. This was all John Couch Adams (1819–1892) needed to look for Neptune.

Dr. Budd died in 1880, the same year scientists isolated the typhus bacillus.

contamination would be if a tiny amount of fecal matter from a sick person somehow got into a well or cistern; it could spread and contaminate that entire water supply. Budd showed that every case of typhoid was connected to a previous illness in another person.

Despite this, many officials believed that typhoid was caused by "miasma," or an accumulation of squalor and bad smells—in short, it was then thought to be caused by being poor.

Little Mary had no idea what Budd had done or that anyone had conflicting opinions. She was born in Ireland in 1869, orphaned early in life, and raised by a grandmother who taught her to cook with whatever foodstuff was available. As a teenager Mary emigrated and, like most Irish lasses coming to America then, took work as a common washerwoman. Since she was trained to run a household and work as a domestic, and since she had dreams, she set her eyes on becoming a cook for some of New York's wealthiest households.

She was very good at her new calling. In the late 1800s well-to-do American families were cultivating an appreciation for fine foods, and New Yorkers were no exception. Mary learned to cook for large groups and fancy dinner parties; she became talented in the making of French cuisine and other European dishes, as well as common meat-and-potatoes fare. Her reputation soared. Her employers knew that Mary would surprise their palates.

What they didn't know was that Mary Mallon was harboring a deadly disease in her intestinal tract, and it was flourishing. No one knows where she might have gotten the typhoid bacteria; some say that she was born with it, having gotten it from her mother's birth canal. The final fact is that Mary herself never got sick from typhoid, but she was a carrier of it.

FAST FACT

Some famous people who died of typhoid (not linked to Mary Mallon): the actor Willard Lewis (1882–1926); First Lady Abigail Adams (1744–1818); and George Washington Gale Ferris Jr. (1859–1896), the inventor of the Ferris wheel.

At first, she probably didn't even know it.

The first family believed to have been stricken with typhoid while employing Mary came down with the disease in 1900, and for reasons that may or may not have been related, Mary lost her job. Because of her reputation, however, it didn't take long for her to find other employment in Manhattan, then another job in the household of a New York lawyer, then in another lawyer's home before she moved up to the Tuxedo Park area. Between 1900 and 1906, she worked for a total of eight families, each of which had members who were sickened with typhoid during Mary's employment with them. History doesn't say if anyone noticed these clusters of illness then, but at any rate, while reports are sketchy on who knew what when, Mary must have added things up on her own because she seemed to job-hop with regular frequency.

In late summer of 1906, things came to a head.

New York banker Charles Henry Warren hired Mary to cook for his family and friends at the end of August of that year at a rented mansion on Long Island. It was a safe area, and the entourage planned to be there for just a few days. In that short time, six of the 11 residents and guests of the home came down with typhoid fever, which alarmed the mansion's owner, who hired an independent expert, who found nothing in the house itself, on the grounds, or in its structure. Ultimately, the soft clams and water were blamed for the outbreak.

Meanwhile, Mary disappeared and went to work for a family on Park Avenue.

But someone else was on the case: investigator and sanitary engineer George Soper (1870–1948) truly enjoyed his job, and he was committed to finding the source of typhoid and other communicable diseases in New York. In the winter of 1907, after having tested, examined, and exhausted all other roads to the possible origin of the deadly disease, he finally met with Mary Mallon. By then a mini outbreak of typhoid had sickened thousands of New Yorkers, and Soper was sure that each case could be traced back to one person.

On the day he finally caught up with Mary, she must've known what was up: he asked for stool and urine samples, and she reportedly lunged at him with a knife. Another public health official, Sara Josephine Baker (1873–1945), was likewise dispatched for the samples and was also sent running away. Finally, five hours later and after no small effort, Baker and a handful of New York's finest caught Mary and escorted her to a hospital, where tests confirmed that, yes, Mary Mallon was a carrier of *Salmonella typhi,* a bacteria that leads to typhoid.

Very much against her will, Mary was immediately sent to live in a small house on North Brother

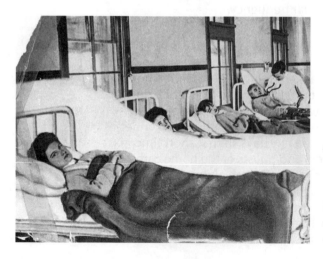

Mary Mallon ("Typhoid Mary") was quarantined for much of her life because she was an unknowing, asymptomatic carrier of the lethal disease.

Island, not far from the Bronx, where she would be supervised so as not to be a danger. While living there, she was treated with methods then thought to help cure typhoid, but she still tested positive the vast majority of the time.

In 1909, she sued the health department, but she didn't win. In 1910, despite continuing to test positive, a new health director forged an agreement with Mary that she could be freed as long as she declined work as a cook. She left the Island with zero plans to follow the agreement, and she went directly to work as "Mary Brown." Almost instantly, typhoid was back in the New York news and Mary was back into custody on North Brother Island.

For years, then, "Typhoid Mary" was confined there, alone, save but for package deliveries and the occasional visitor who probably barely touched her. After an illness in 1932, she was taken to a hospital, where she lived for six years until she died in November 1938.

Shortly after her status as a carrier was confirmed, Mary was offered a cure—the removal of her gallbladder—but she strongly refused. After her death, postmortem tests on her gallstones showed typhoid bacillus, leading to questions of whether she could have been cured had she acquiesced to surgery.

For the last 38 years of her life, she refused to take responsibility for the typhoid deaths of at least three and as many as fifty people. To anyone who'd listen, Mary insisted that she had done absolutely nothing wrong.

FAST FACT

In the 1830s, William Whewell (1794–1866) invented the word "scientist" when writing about Mary Somerville. Up until then, astrologists, biologists, and the like were called "men of science," which obviously didn't apply to a woman like Mary.

But she did: she failed to wash her hands often, which would have avoided this whole thing.

Plants and Animals: Please Don't Slay the Dragons

Though you'll find dragons all over in mythology and legends, and though you might wish they were real, the truth is that dragons don't exist. Or, well, they *do*, but in name only.

Komodo dragons—also known as Komodo monitors—are members of the monitor family and are the largest and heaviest species of lizard currently alive. These big boys are not exactly pet-sized: Komodo dragons can grow to a length of 10 feet (3 meters) and can weigh more than 300 pounds (136 kilograms) at the extreme.

When a Komodo dragon hunts—which they do almost constantly—they don't stalk prey like many hunting creatures. Komodos will wait patiently for hours while they "taste" the air with their long, forked tongues. As soon as prey wanders close enough, the dragon pounces at speeds of up to 13 miles (almost 21 kilometers) per hour and latches onto the meal with huge serrated and curved teeth.

The largest species of lizard, the Komodo dragon is a monitor lizard native to Indonesia. It can grow to almost 10 feet (3 meters) in length and weigh 150 pounds (70 kilograms).

The unfortunate creature that ends up on the tooth-end of a Komodo dragon doesn't have a chance. If the dragon brings the prey down alive, his lower jaw unhinges and his stomach expands, making it possible for the dragon to devour living prey quickly, in large chunks or whole. A Komodo dragon can consume up to 80 percent of its own body weight in one meal, and if it needs to flee, it can regurgitate everything to make it easier to escape.

If the prey manages to escape, that's okay too (for the dragon). Because his mouth is full of bacteria (up to 50 different kinds), anticoagulants, and highly venomous sacs, the dragon's bite will cripple and eventually kill the prey all the same, usually in a matter of days. The dragon then uses his awesome scenting ability to find the carcass and eat that. And if he doesn't find the dead prey or if another dragon eats it first, that's never a problem. Komodo dragons aren't fussy; they'll eat any kind of meat, including human, and carrion or roadkill, which they can smell more than two miles (more than three kilometers) away. They'll even eat another Komodo dragon, if there's nothing else in the pantry.

The nice thing is that a dragon will share his kill with another dragon—and if they do fight, as they will in competition during their spring-and-summer mating period, one dragon's venom and bacteria won't kill his foe. Scientists don't know why yet.

If you haven't figured it out yet, you don't want to mess with a Komodo dragon. They don't often attack humans, but they absolutely *will*, and their bite could kill you within a few hours unless you get *immediate*-immediate help. You may be able to outrun one, but not for long; zigzagging might work, but it might not. Komodo dragons make *really* lousy pets.

But please don't slay these dragons. Komodo dragons are mostly found on a small handful

> The black spiny-tailed iguana, native to Mexico and South America, is also sometimes called the Wish Willy.

of islands in Indonesia and in a few zoos that are equipped to care for them. In the wild, there are fewer than 6,000 of them, and for that, the International Union for Conservation of Nature has listed the Komodo dragon as threatened.

Human Body: This and That and the Other Thing

If you're an average adult, you'll lose many billions of cells each day to cell death. No need for a funeral; they're replaced as quickly as they die by brand-new cells, and you don't even have to think about it.

Let's start by saying that many adult humans don't drink enough liquids. But this eight-glasses-of-water-a-day stuff isn't entirely true—first of all, because it doesn't take into consideration all the *other* things we drink. Tea, juice, even soda will hydrate a person, even just a little bit; sports drinks are also good. Another thing: it's rare for a modern diet not to offer hydration; in fact, about half our water intake comes from the meals we eat.

You don't have any muscles in your fingers. What moves them are the tendons and muscles at the base of your fingers, in your hand, wrist, and arm.

You don't have to literally drink eight glasses of water a day to keep hydrated because the recommended amount can also be found in other liquids you drink as well as foods.

Here's something you can't buy online: Purkinje fibers. The good news is that you probably already have all you need: Purkinje fibers are found inside your heart, and their sole job is to conduct an electrical pulse that helps the heart contract and release to ensure blood flow throughout your body

You are never alone: your eyelashes and eyebrows are colonized by Demodex mites. These tiny arachnids are most active at night, and you should be glad: they dine on dead skin cells and an overabundance of oils on your face, which helps your skin. But don't be alarmed: most people have Demodex mites; they're only a worry if their population runs out of control.

Fun experiment: Grab a bathroom scale and squeeze the weighing part with your hands without making a face. Now do it, grimace like crazy, and you'll see that you were able to squeeze harder when you make that "grrrrrrr" face. Science says that you're stronger when you grimace because of a brain cortex connection between hand muscles and facial muscles, which means that clenching your jaw increases the strength of your grip.

The first medical school in the American South was the College of Medicine in Baltimore, Maryland. It was started in 1807.

You have several different kinds of fat on your body: brown fat (which works as fuel and to keep you warm), white fat (where your extra calories are stored), essential fat (that which is needed to live), beige fat (which is like a cross between brown fat and white fat), and pink fat (associated with lactation). They are located subcutaneously (found just under your skin, mostly acting as cushion) and viscerally (encircling your innermost organs; this is the fat you lose when you diet).

Science anxiety is a real thing. It's the diagnosis you'll get when you simply cannot cope with any scientific subject matter.

Plants and Animals: Menagerie, Part 2

Male ants—whose only purpose in life is to mate with the queen—come from unfertilized eggs. How this happens is still a scientific mystery.

When someone is said to be blind as a bat, that means they can't see a thing. But guess what? Bats aren't blind. Nope, they very clearly have eyes and they can see just fine, but they *prefer* to use their sense of echolocation to get around since a lot of the flying they do is nocturnal. Echolocation uses a series of quick, high-pitched yips and clicks that hit objects auditorily and bounce back to the bat's ears, giving the animal's brain a sense of where the object is.

Cats' spines have the ability to twist and turn in order to re-orient their bodies so that they can usually—*but not always*—land on their feet when they fall. If the distance they've fallen isn't far enough and there is not enough time for them to twist their bodies, or if the cat's too fat to be able to turn as much as it needs to, the kitty could be injured or killed. So don't even think about testing it.

Some of that lovely white sand on mainland beaches is quartz. And some of it is basically parrotfish poop. As a meal, the parrotfish nibbles on coral and algae on the edges of reef rocks and, in the process, swallows some of

A display of the Argentino-saurus at a Padua, Italy, exhibit in 2017 provides an idea of just how gigantic this dinosaur was.

the reef material. The fish's stomach then breaks down the skeletal remains of the coral (calcium carbonate) and excretes what's left, and you'll remember that the next time you're on vacation, won't you?

Have scientists already discovered the largest dinosaur? Possibly. Based on fossil pieces discovered in India in 2023, the Bruhathkayosaurus was a sauropod speculated to reach a length of 115 feet (35 meters) and weighing up to 170 tons. As with most early discoveries with incomplete evidence, the scope of this amazing beast has not yet been verified to the satisfaction of all scientists. If this isn't confirmed, then the biggest dinosaur would remain the Argentinosaurus, which was discovered in 1993 and was 131 feet (40 meters) in length.

Dinosaur fossils have been discovered on each of the seven continents.

Poor Finny. You don't think he's very smart, being a goldfish and all. You think he's only got a three-second memory, yet you've trained him to come to the surface of his bowl for food. If he only had a three-second memory, he wouldn't be able to be trained, right?

There are several differences between an herb and a spice. An herb is generally the leafy part of the plant; a spice can come from the root, stem, flower, or bark. Spices have stronger flavor than herbs and are generally dried. Herbs can be dried or fresh.

Talk about a bad case of constipation: some bears don't eat, drink, urinate, or defecate for the entire six months or more of hibernation.

Human Body: To Save a Million Kids

Forget cheesy movie monsters. Forget horror novels. In the 1940s, '50s, and '60s, the thing that put real fear into the hearts of people around the world was microscopic in size.

Archaeological evidence suggests that *Poliomyelitis* (the polio virus) was around at least a few thousand years ago, but it wasn't until 1789 that British doctor Michael Underwood (1736–1820) formally described the disease: it displays an innocent onset with what we might call flu-like symptoms that seemed to go away after a few days—but sometimes it culminates in cases of terrifying partial or total paralysis. Underwood noticed that the disease was particularly hard on children, but adults weren't always left unaffected.

Sixty years later in 1840, German orthopedic doctor Jakob Heine (1800–1879) studied the disease and put forth the idea that it wasn't caused by humors or vapors or anything except contagion, passed between people directly.

In the late 1800s, what was by then known as "infantile paralysis" (because of the age of most patients, though many were well past infancy) had spread to the United States and caused a huge outbreak that started in Rutledge, Vermont. Documented symptoms often started off relatively mild, with a headache, sore throat, and fever, which sometimes led to stiffness and paralysis in certain

When Dr. Salk's polio vaccine was proven effective, the story made headlines all over the world.

parts of the face or body. In worst-case scenarios, patients were left with total paralysis, or they died.

What baffled doctors was the disease's communicability. They didn't realize then that some sufferers were asymptomatic, which complicated things until 1907, when Swedish doctor Ivar Wickman (1872–1914) confirmed Heine's idea about contagion, that infantile paralysis could be found in people who had no symptoms. This was important, but it was no cure.

A year later, there was a breakthrough.

Obviously, with a disease like infantile paralysis around, scientists everywhere were racing to find a cause and, they hoped, a cure. In 1908, Viennese doctors Karl Landsteiner (1868–1943) and Erwin Popper (1879–1955) experimented on monkeys with spinal fluid from a nine-year-old deceased infantile paralysis victim, and they determined that the disease was caused by a virus, not bacteria.

Further experiments by other learned scientists confirmed the awful truth: polio is very highly contagious, it lives in the throat and gut, and it spreads generally through contaminated fecal matter. It could sicken its victims quickly, or it might not present with any symptoms at all, leaving the victim to continue to spread polio to others for weeks. Some victims fell ill to post-polio syndrome, which caused the illness to return to some degree. Some people fell sick with polio and healed with no ill effects; in other homes, entire families were affected. Humans are the only natural hosts of polio. When a victim got polio in those early days, there was no one specific way to treat them.

Obviously, hand-washing and general good hygiene were important to stop the

In the early 1950s, polio caused tens of thousands of cases of paralysis per year in the United States. In 1952 alone, there were a reported 21,000 cases of paralytic polio.

spread of polio, but it wasn't enough; it could also spread through droplets from a sneeze or cough. It could potentially come from contaminated food or water. It could come from something as simple as changing a diaper. And victims weren't necessarily found in tenements or squalid conditions: in 1921, Franklin Delano Roosevelt (1882–1945) was stricken, possibly at a Boy Scout camp near his family's summer cottage.

By 1931, Australian physicians Frank M. Burnet (1899–1985) and Jean Macnamara (1899–1968) had determined that there was more than just one type of the poliovirus and that victims could be stricken with more than one type. Twenty years later, at the height of what was surely a time of terror for parents everywhere, virologists David Bodian (1910–1992) and Isabel Morgan (1911–1996) were able to determine that there were three types of the poliovirus. Their work opened a door on the possibility of a vaccine.

FAST FACT

During his second term, President Roosevelt decided to ask Americans' help to fight polio— but it was 1938, and because of the Great Depression, money was still pretty tight in most households. At one fundraising event, singer Eddie Cantor (1892–1964) suggested, as a joke, that Americans should send Roosevelt dimes. It was no  joking matter to everyday citizens, who sent scads of ten-cent pieces to the White House, prompting the National Foundation for Infantile Paralysis to change its name to the March of Dimes. The effort is also credited for putting Roosevelt's face on the modern ten-cent coin.

This was a big deal, with a clear conclusion: that vaccine was needed, and soon.

In 1948, virologists Thomas Huckle Weller (1915–2008) and John Franklin Enders (1897–1985) and pediatrician Frederick Chapman Robbins (1916–2003) were able to grow live polio viruses in live human cells in test tubes, which paved the way for further research, and for which they received Nobel Prizes in 1954. Meanwhile, the number of cases of polio rose.…

In 2020, nearly 1,900 paralytic polio cases were reported around the world. As of the time of this writing, a tiny handful of people stricken with polio in the 1950s still rely on an iron lung (a large tube they literally live in) to help them breathe.

One year later—whew!—Dr. Jonas Salk (1914–1995) finally presented the world with a vaccine to prevent polio. Among the first to get the vaccine were Salk's wife and his sons, followed by millions of kids in tens of thousands of pediatricians' offices and other venues in 90 countries around the world. Kids who hated getting shots rejoiced in 1961 when Albert Sabin (1906–1993) released the live oral polio inoculation, which was generally given to kids on a sugar cube to mask the taste.

Dr. Jonas Salk was lauded worldwide for his polio vaccine, receiving many awards, including the Presidential Medal of Freedom, for his work. He later founded the Salk Institute for Biological Studies in La Jolla, California.

> **FAST FACT**
>
> According to singer Lou Reed (1942–2013), the song "Save the Last Dance for Me" was written by singer-songwriter Doc Pomus (1925–1991), a polio survivor who used a wheelchair and occasionally crutches. On the day Pomus got married, he sat in his wheelchair and watched his actress-dancer bride, Willi Burke (1933–), as she danced with their family and guests, and the song simply came together.

Dr. Salk did not patent his vaccine, nor did he ask for any profit from it because he reportedly wanted it distributed as far and wide as possible to save as many lives as it could. He didn't stop there: before his death in the 1990s, this very good man went to work in search of a vaccine for AIDS.

Polio continued to plague underdeveloped nations well into the 1970s, but by the turn of this century it was considered to be all but eradicated, except in a small handful of countries.

- *The last case* of type 2 polio was in India in 1999.
- *The last wild* case of type 3 polio was in 2012.
- *In 2020, about* 700 cases of polio were reported worldwide.
- *One year later,* the wild poliovirus was discovered in wastewater in New York City and London, and there was a resurgence of the disease in Mozambique.

Plants and Animals: Playing Possum

Chances are you've seen the meme about how beneficial possums are to have in your yard. Turns out that there's a whole lot more you should know about the critters.

The opossums you are likely most familiar with were once found in North America before becoming extinct some 25 million years ago. About 3 million years ago, they came back to North America by land bridge from South America, where they never went extinct.

The animals are marsupials; in fact, they're members of the largest group of marsupials in the Western Hemisphere. There are 120 species of opossum, but just one, the Virginia opossum, is found naturally in the United States and Canada.

The only difference between an opossum and a possum is the missing "o." Otherwise, we're talking about the exact same creature. In case you're curious, the name probably comes from the Powhatan language, and its use goes back at least 400 years.

Although the possum uses his fierce hiss-growl when disturbed, he's really a nice guy, in the end. Possums are not usually aggressive; the large open-mouthed, fierce-looking snarl is only a warning to stay away. Oh, and they will also scream, if they're worried or disturbed. Yes, they could bite you with their *fifty* teeth in defense, but it's rather rare that one would.

While opossums only live in North America, they have distant cousins that are also called possums in Australia. Possums there include 27 species (including related "gliders").

As for size, a possum grows to be about the size of a pet cat (although they can occasionally get much bigger), with a pointy nose that possesses well more than twice the olfactory receptors you have in your nose. Its ears turn dark as a possum ages; a possum's ears are like human fingerprints, in that each one has a unique-to-the-individual pattern of black markings. Possum ears and tails are very prone to frostbite because both are considered to be hairless.

A possum's body hair overall is wiry and feels a little like scrub brush bristles.

Even if you don't find anything to admire about a possum, you do have to give them props for their dexterous little feet. The front feet largely resemble human hands, in that they are used for grabbing, holding, and climbing; also, like human hands, possums have palm- and fingerprints. The back feet sport four clawed toes and one opposable toe without a claw.

Despite what pop-culture cartoons want you to believe, an adult possum's tail isn't strong enough for it to hang upside down (although a baby possum's tails will hold the infant for a little while). Adults have longer prehensile tails (unless they're shortened by frostbite) that are scaly and mostly devoid of hair.

You never have to kindly feed your neighborhood possum. Opossums are opportunistic feeders and omnivores, meaning that they'll eat small mammals, bugs, worms, seeds, fruit and nuts, but they'll also raid any dog or cat food you happen to leave out, and they'll dine on garbage, if it's on

Opossums are biologically interesting animals. They are the only marsupial species in the North America, and they have more teeth (50) than any mammal on the continent.

the menu for the day. Sadly, they may take bird eggs or slow-moving poultry, if they can. As for the rumor that possums eat ticks, the jury's still out: some scientists say yea, some say nay.

> If you're up for some fun, visit the Possum Monument in Wausau, Florida. It's there to laud the "magnificent" possum in North America and all that it has to offer humans on the continent.

Daddy possum has a forked penis; fortunately, Mommy possum has a forked vagina to accommodate mating; generally speaking, Mommy possum produces two litters of babies per year. Gestation is roughly two weeks, at the end of which up to 25 pups must find their own way to their mother's pouch, where around 15 of them will attach to a teat and do little more than nurse and grow for around two months. It's true that the female opossum will carry her clinging young around for safety's sake for awhile when they're very small. It's also true, alas, that it's very easy for baby possums to get detached, lost, or forgotten in a hurry.

Though they like to be up and about all night long, possums are not party animals. They prefer to live and forage alone, although small families of them will live in burrows that another animal has dug and abandoned. Many species of possum also like to spend time in trees, and will climb just as readily as they'll want to go underground. Just FYI, you'll want to be sure that your new pal doesn't get stuck in a barrel or bin that he can't get out of, because cold weather and icy containers are not good for a possum.

One of the most famous—and perhaps most amusing—things a possum will do is ... well, he'll "play possum," which is to say that he'll act as if he's dead when he feels at all unsafe and the hiss

> Possum fur, specifically the possum's undercoat, glows pink in ultraviolet light.

doesn't work. Though it's an involuntary response to a perceived threat, his act is Oscar-worthy because it comes complete with a nasty smell and a rigor mortis-like stiffness. His fake little carcass can be picked up, and he won't give up the ruse—perhaps for as long as a few hours. When he's back to normal, the possum will just waddle off to live another day.

Speaking of living, some larger species of possums seem to be immune to snake venom. Science is looking into this.

After all this, it might be hard to believe that the Virginia possum was hunted in the not-so-distant past and was said to be a favorite dish of author Mark Twain (1835–1910) and President William Taft (1857–1930). Also not so long ago, the critters were farmed like cattle and chickens.

In early 1909, after an Atlanta banquet at which President Taft was feted, the great man was presented with a small stuffed opossum that was supposed to be the next big fad after the Teddy Bear, named after President Teddy Roosevelt (1858–1919). Creators called it the Billy Possum, and thousands were made. Though it sounds so darn cute you almost can't stand it, the Billy Possum was doing little more than just playing dead by the end of the year.

FAST FACT

Possums + Presidents = Passion? In 1892, President Benjamin Harrison (1833–1901) specifically requested a pair of possums from a constituent; it was never said whether he wanted them for pets or dinner. President Richard Nixon (1913–1994) was a member of the Possum Growers and Breeders Association of America in the early 1970s; Jimmy Carter (1924–) joined sometime later.

FAST FACT

Scientists are torn between saying that preference for smells is something we're born with or something we learn. Others say that it's a little of both: by 10 weeks' gestation, a fetus has the ability to smell; as a child, they will learn to associate some smells with comfort and others with danger.

Human Body: This Really Stinks!

Take a deep breath.... Once upon a time, scientists didn't think your sense of smell was all that important. The belief was that your eyes and your ears did most of the work of your senses. Now they know better: your nose is a key to many social and psychological impacts on your well-being. It helps you identify people, places and things, and it keeps you safe. It's thought that your olfactory system can also affect your personality.

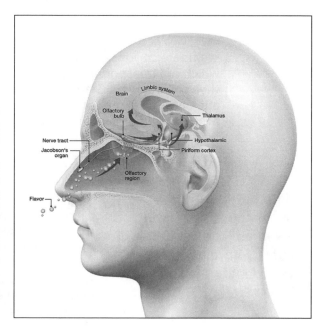

The anatomy involved in how we smell is more involved than you might think. Once perceived by the olfactory receptor cells, smells can stimulate emotions, memories, and even behavior.

While a talented hound dog can find a person by the scent they leave behind on a footprint, your pathetic little nose has an olfactory area at the back of your nose that's less than an inch square and holds around 50 million receptor cells. (Fido has up to 300 million receptor cells.) You have a few hundred genes that support your ability to sniff; most animals have more than a thousand such genes. It's enough to make you want to snort.

Think about this, though: once humans started walking upright, we didn't *need* all those receptors and genes because our noses were sometimes just out of range of most scents, which tend to hover closer to the ground. Today, 2 percent of all humans suffer from disorders of the olfactory system; men have more smell disorders than do women, and the sense of smell for both sexes tends to diminish with age. And yet scientists say that your sense of smell is pretty good, all things considered.

But your nose isn't just an air-exchanging feature hanging on your face. It has *work* to do.

Take, for instance, the ability of a scent to evoke memory.

Two of the most familiar, most memorable scents are crayons and freshly mown grass, both of which can transport a lot of American adults back to their childhoods. If Grandpa used a certain kind of aftershave, the smell of it can remind you of him—even decades after you last sniffed it. Favorite foods, best-loved places—the scent of any of them works on your brain to bring back mental images.

Nine out of 10 mothers of newborns, after having spent as little as 10 minutes with their babies, can identify their infant by smell.

Or maybe those scents cause a sort of overreaction in the olfactory system. In that case, you might be a super smeller, the casual term for someone who suffers from hyperosmia or an elevated sensitivity to scents.

FAST FACT

Did you know that some scents have names? Bibliosmia is the smell of old books; it comes from the breakdown of the book's paper. Petrichor is the smell of rain, generally the first rain on dry ground.

Imagine being able to smell if someone has cancer or multiple sclerosis. Or being able to detect minuscule amounts of chemicals in a substance. Or imagine helping researchers to rescue scents that are in danger of going extinct (yes, that's a thing). Those are things a super smeller can do, but that's not all.

Having hyperosmia is somewhat rare and may be linked to certain diseases such as Lyme disease and migraines, fluid shifts in the body, hormone fluctuations (especially in pregnancy), and some prescriptions. Genetic mutations may also be at fault. There may be other indicators; researchers aren't entirely sure yet what causes hyperosmia, but they do know that small differences in an individual's brain can point to possible super smeller potential. Like many things, it's best if it's checked out by a doctor.

If you are a genuine, permanent super smeller, you might be in luck: there are jobs waiting for someone with a refined nose like yours. Perfume makers need your talents to make scents that bring to mind nostalgia, innocence, or lust. Researchers need you to figure out why certain creatures are attracted to certain human smells to bite us or flee from us. Foodies need help making sure that dinner smells as tasty as it looks, and drinkers need you when they pop a cork on the wine. And, of course, deodorant manufacturers need armpit and foot sniffers to make their product work best.

Intact billy goats (meaning those that have not been neutered) have a reputation for smelling nasty.

The reason is their intoxicating cologne: male goats will urinate on themselves in several places—legs, belly, beard, and head—to maintain their musky smell and their appeal to the nanny goats.

Earth Science: These Things Really ROCK!

A rock is a rock is a rock, right? You know that there are three main types of them (igneous, metamorphic, and sedimentary, in case you need a refresher), but otherwise, rocks are basically inert things of different sizes.

Trovants

Or not. Sometimes rocks seem to do things that might make you think they're alive.

Deep in the hills of Romania lies the small town of Costesti. It's home to around 11,000 people and several "living" rocks called trovants. The folks in Costesti say the trovants can move about, they can grow, and they can reproduce at will. They swear that the *life* of a trovant is not just folklore, and science agrees.

Trovants come in all sizes, from thumb-size rocks to boulders that can reach up to 15 feet (4.6 meters) in height with weights that are about average, commensurate to their size. If you simply spotted a bunch of them, you would notice that they tend to be roundish or that they look like fat Frisbees lying in the sun; some of them might look like big fried eggs. Locals used to claim that trovants were brought by aliens or were dinosaur eggs, but they're rocks—that's it. Otherwise, there's not much remarkable about them.

And yet, there is.

When it rains, the calcium carbonate in raindrops mixes with limestone to make a form of cement that oozes from trovants. This causes grains of sand to adhere, layer upon layer, which makes the rocks grow very, very slowly—about 2 inches (5 centimeters) every 10 centuries. The concrete substance tends to keep the stones roundish, but if it bubbles off to one side for any number of reasons, the side-bubble can get too heavy and fall off—which seems to be how trovants "reproduce."

Even weirder, some say that trovants can seem to move, albeit very slowly. Science says that trovants rest on what may have been ocean floor once upon a time; fossilized shells have been discovered deep inside a trovant. Any so-called movement has gone without eyewitnesses, so questions remain. One explanation for movement, if any, seems to be the heating and cooling of the soil beneath the trovants, but that's not a for-sure. Rains and loose sand might also be to blame. (Keep reading, there's more below on traveling rocks.)

You can visit trovants if you're heading for Romania; but just know that trovants are protected by UNESCO (the United Nations Educational, Scientific and Cultural Organization).

The Sailing Stones of Death Valley

They say that a rolling stone gathers no moss, but there's nothing in that old saw about rocks like the ones in Death Valley National Park: those rocks

How the sailing stones moved across the terrain in Death Valley was a spooky mystery for many years, but scientists eventually figured it out in 2014.

seem to crawl across a dry, sandy area called Racetrack Playa. The thing is, nobody's ever actually *seen* it happen, but that's okay: the stones leave trails across the dry sand bed, and there's your evidence.

Originally, the rocks came from the mountains that surround the playa (a word denoting a flat, arid field); we know this because the rocks on the field are made of the same syenite and dolomite that the mountains contain. So the rocks loosen and fall off the mountains due to erosion, roll into the playa, and start to shimmy.

These aren't little pebbles either. They're not the kind of rocks you can kick: some of the sailing stones weigh upwards of several hundred pounds. Should you decide that you don't like where they are, you should know that most of the sailing stones definitely aren't going anywhere without a few strong humans and a backhoe—except when conditions are right. When that happens, the trails behind displaced stones indicate that an individual rock can move seemingly *all by itself* for up to 300 yards (274 meters) at a speed of up to 16 miles (nearly 26 kilometers) per hour. That's faster than the average human can walk.

The good news is that scientists using a camera figured things out: though Death Valley's upper temperatures are notoriously stratospheric (some of the highest temperatures in the world have been recorded there), it can get surprisingly cold there in the winter—cold enough for water to freeze. Yep, they didn't know it before about 2014, but there's a shallow pond in the playa that's less than an inch (less than three centimeters) of water deep, when there *is* water. As the winter temperature drops, the pond freezes, causing a very thin sheet of ice to form in the sand around the rocks. As the daily temperature rises, the ice melts and floats, building up behind the rocks. The wind blows, the ice blockade moves the rocks in front of it, trails are left, and we can be awestruck.

"Singing" Sand

Okay, sand isn't exactly a rock—it's a former rock, technically speaking—which makes the fact that it "sings" even more phenomenal.

It's said that Marco Polo heard singing sand in the deserts he traveled, and he thought it was caused by evil spirits of the desert or some sort of ethereal, out-of-tune orchestra. He was way off in his hypothesis: though they don't know the whole story yet, scientists know that singing sand is caused by moving air.

Specifically, as sand piles up in dunes, it creates a sort of tipping point at the top, or an about-to-be overflow. When conditions are right—the right shape of the sand grains, the right humidity, the right amount of silica—the sand will start to fall, which will make it vibrate, causing the air to vibrate too, which makes the sound of a low moan or a million voices humming softly.

FAST FACT

If you've ever had to check a geologic map, you have British geologist William Smith (1769–1839) to thank. In 1815, he developed the science of stratigraphy, which studies and determines the strata of the Earth. Some of the names of the strata that we use on maps today were bestowed by Smith some two centuries ago.

This low frequency, around 95 to 105 hertz, lasts up to 15 minutes and can be heard for miles. If you want to hear them, you can visit the Great Sand Dunes National Park in Colorado or any of the other places on Earth where sand sings.

Space Science: Smallest to Biggest (and in Between)

Want to feel very, very small? Or perhaps very, very tall? Then read on....

- *Jupiter is the* largest planet in our solar system, with a radius of more than 43,000 miles (almost 69,900 kilometers). The planet that's closest to the Sun, Mercury, is the smallest, with a radius of just over 1,500 miles (2,400 kilometers). Compare these to Earth, at just a little less than 4,000 miles radius (6,300 kilometers).

- *It's a little* difficult to determine superlatives with stars, since science gets more knowledgeable about them every year. At the time of this writing, we do know that UY Scuti is one of the largest, with a radius that's about 1,700 times larger than the radius of the Sun. Conversely, Vela Pulsar is a mere 12 miles (roughly 19 kilometers) in diameter. You could definitely walk that.

- *Comet Bernardinelli-Bernstein,* also known as C/2014 UN271, is the largest comet ever seen. It's more than 70

The James Webb Space Telescope has taken some fantastically detailed photos of our solar system's largest planet, Jupiter.

miles wide (almost 113 kilometers) and it orbits the Sun. The smallest (so far) is Comet Hartley 2, or 103P/Hartley, which is not quite a mile (1.6 kilometers) wide.

So far, the star that's farthest away from Earth (but still in our galaxy) is Earendel, which is 28 billion light years away.

Here's the order of this branch of science's terms: diminishing in size, it starts with the universe, then galaxy, solar system, star, planet, moon, asteroid.

The largest telescope on Earth is the Gran Telescopio Canarias at the Roque de los Muchachos Observatory near Spain, at just a hair over 34 feet (nearly 11 meters) in diameter. That's the word at the time of this writing; astronomers are constantly building bigger and better ways to see the stars.

Neptune has 14 moons. Saturn has 146 known moons, and there may be more that are embedded in its rings. Neither Mercury nor Venus has any moons.

While you might think Mercury, the planet closest to the Sun, would be the hottest planet, it's not. Venus, with a large atmospheric blanket around it, is the hottest planet in our solar system at somewhere around 900°F (482°C). Mercury's temp, by the way, is a mere 800°F (426°C).

Here's another surprise: though it's not at the tail end of the planetary lineup, Uranus was the planet with the coldest recorded temperature at –371°F (–224°C). Its distance from

FAST FACT

On March 13, 1781, astronomist William Herschel noted a slight something in the constellation Gemini. After he realized that it was a planet, he hoped to name it after George III of England (1738–1820), but other astronomers stepped in with their mythological traditions and named the planet Uranus.

the Sun is not the reason: eons ago, a crash knocked Uranus sideways, and it's been revolving around the Sun sideways ever since.

 Galileo Galilei (1564–1642) discovered more than 400 years ago that Saturn has rings surrounding it. Science has learned since then that there are up to 14 rings, and they are made of rocks and ice of sizes that range from tiny to gigantic. Some of the rings may have formed recently, but it's more likely that they formed long ago, either from broken bits of moon or from flotsam left over when Saturn was formed.

Notable Names: Who Was Harriet Cole?

Though the Emancipation Proclamation was signed in 1863, the life of a black woman in America was far from good in the years afterward. Black *men* had gained the right to vote through the 15th Amendment in 1870; black women didn't get that right for decades. In the late 1800s, most former slave women took jobs as sharecroppers, domestics, servants, cleaners, and janitorial-type workers. History makes note of a sad few stories, but the details of the lives (and deaths) of most former slaves weren't recorded.

That was where Harriet Cole ended up, as the story goes.

But first, let's say that there's so much unknown about Harriet Cole, beginning with who she was.

Census records exist for Harriet Cole, who worked in Philadelphia as a domestic, but we have to remember that the work was common, as likely was that simple name. At any rate, the Harriet Cole on the census seems to be the same Harriet Cole for whom an 1888 death certificate was filed that indicates her age as roughly 35 to 40 years old, and that records Drexel's College of Medicine (Philadelphia's Homeopathic Hahnemann Medical College) as her "place of burial."

Both documents line up with the presence of Dr. Rufus B. Weaver (1841–1936), his work, and the fact that he received a cadaver that records indicate was Harriet Cole's dead body.

Weaver dissected cadavers; for an unknown time, Harriet cleaned up after him.

It's unrecorded whether the two were friends. Did they make small talk together? Did Weaver treat his employee with kindness? There must have been at least some level of respect because, it seems, Har-

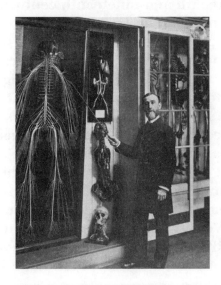

Dr. Rufus B. Weaver presents the nervous system of Harriet Cole, which he carefully removed from her body over a period of months, preserving every fine strand perfectly.

riet decided to leave her body to science—specifically, to Dr. Weaver because, it seems, she understood the difficulty he had in procuring cadavers for study. There is no paperwork to that end, but at that time in history there might not have been much, if any, documentation for a black woman. Either way, she might or might not have known what he had in mind for her remains.

Weaver, born in Gettysburg, Pennsylvania, stepped into his job quite naturally.

At the end of the Civil War, his father, Samuel, was a photographer and was tasked with helping to identify the Confederate dead. When he was killed in a railroad accident, the job fell to the younger Weaver, possibly because he'd become an anatomist, which was a new category of scientific study. Once his work with the Confederacy ended, Weaver landed at Hahnemann in 1879. Harriet, it appears, was already working there.

You have to wonder if he even noticed her at first. His work then was his focus—he was an anat-

FAST FACT

In the eighteenth and nineteenth centuries, dissection was considered to be so abhorrent that it was part of the punishment for prisoners. In order to procure enough fresh corpses for medical students to learn from, a whole industry sprang up. Body snatchers stole newly buried people right from their graves and sold the bodies to medical schools because few people willingly volunteered their remains for dissection. It's been suggested that Harriet Cole likely didn't *willingly* donate her body but that she was a victim of racist perspectives of the time. It's true that her willingness can't be confirmed. It's also true that Weaver's reputation was squeaky-clean during his lifetime, so draw your own conclusions.

omist, after all—and he wasted no time in collecting medical specimens while dissecting corpses with his students. He indicated to a colleague that he struggled to teach his students about the nerves and brain without a good example for them to see, and he was scolded for his dream of dissecting one from a human donor. His idea was considered to be folly.

And then Harriet died.

We know that she died of tuberculosis, but we don't know exactly where, or even exactly when, and we don't know how she arrived in Weaver's laboratory. Did a loved one bring her, or did someone fetch her body? Or was she—as is suspected today—coerced into the arrangement because she was an uneducated former slave, and he was a wealthy white man? History doesn't say, but the first thing Weaver did was to inject Harriet's body with zinc chloride, the usual disinfectant of the time. He then laid her in a tub for an unspecified period.

Details of the next few months are sketchy because Weaver kept few notes. We don't know what solution he used to preserve Harriet's body, or how he managed to untangle each nerve from the top of her head to her toes without breaking any of the hair-thickness strands. The spinal column must've been particular thorny to do because of the narrow width of the bundle of nerves. No one knows for sure how long the process took, although it's suggested that the *dissection* might have taken anywhere from five to almost ten months, start to finish, not including the many processes he used to be able to display Harriet's nervous system.

When he was finished, Weaver was lauded by colleagues and students. He'd completed the project in time for Harriet's nervous system to

Britain's Prince William says he doesn't get nervous when he gives speeches. He simply doesn't wear his contact lenses, so he can't see the audience that's watching him.

be displayed at the Chicago World's Fair. After count-less fairgoers viewed the marvel of the human nerv-ous system, it returned to Philadelphia, to Weaver's museum, where Harriet Cole sits today.

Human Body: Smile Big!

Everybody stand together. Okay, say "CHEEEESE" … Let's begin at the beginning: as long as there have been humans, humans have had teeth. Still, we have evolved: our jaws seven million years ago would have been a lot like those of today's chimp, filled with thinly enameled teeth arranged in U-shaped rows with room for long ca-nine fangs and no chins to speak of. Today, our teeth have thicker enamel and are in a semi-oval shape, our canine teeth are much shorter, and—yay!—we have chins.

Back then, our jaws were bigger, with plenty of room for what we call wisdom teeth. Big molars were no problem either. Still, like us, prehistoric hu-mans had 20 teeth as small children and 32 adult teeth, including a pair of nice-sized canine teeth. The fact that modern teeth are smaller reflects a downward evolutionary trend for millions of years. Fun fact: some modern humans don't naturally de-velop wisdom teeth, which is evolution in action.

The oldest dentures ever discovered were found in Mexico and are thought to be about 4,500 years old. They were made of animal teeth, which were, along with human teeth, the usual material for a long while.

The good news for an-cient people was that they probably rarely needed a den-tist, although archaeologists *have* found evidence of treat-ments for tooth decay in 14,000-year-old teeth—with-out Novocain, mind you—and ancient skulls have also been discovered with abscessed teeth. Overall, though, while we floss, brush, rinse, and see a dentist often, ancient people probably had healthier teeth because of their diet of meat,

A human skull from 90,000 years ago (left) shows a much larger mandible compared to a modern human. Diet back then required more chewing,

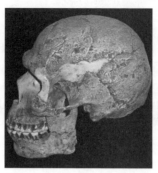
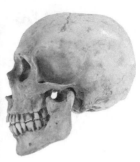

plants, and nuts, versus ours that is heavy in grains and sugar.

But here's the best part about ancient teeth: there are *so many of them* that have been found and are available for study! For every bit of bone or cranium that an archaeologist might find, there are proportionately hundreds more fossilized teeth at an average dig site, in part because teeth are stronger than bone in life and in death.

The secrets that ancient teeth can tell are many: by examining them for details that are still on the enamel, even after all these centuries, scientists can determine the approximate age of the mouth those teeth were in, the things those teeth chomped on, how the owner of those teeth migrated and where, the diseases their owners suffered from, what they hunted, and the amount of time spent indoors, outside, and in clothes. We know from ancient teeth, for instance, that our Neanderthal ancestors were often plagued by malnutrition, and that the bacteria in the mouths of the general human population changed when humans became agrarian.

When today's infant is around 6 months old, parents may notice that their teeth are

> Scientists have lots of teeth at their disposal for study, but they know surprisingly little about how our ancient ancestors' teeth erupted, were cast off, or grew when they were youngsters.

FAST FACT

An infant is born with all its primary teeth ("baby teeth") already in its skull but out of view in its jaws. Baby teeth begin developing a few weeks after conception. About 1 out of every 2,000 babies is born with a tooth or two.

erupting from their gums, generally starting with the upper and lower front teeth (incisors). Primary molars join those teeth roughly four to six months later, followed by the canine teeth and the large molars. These teeth differ from adult teeth in both color—babies' teeth are whiter than their parents' chompers—and shape, being more squarish and smaller.

Here's how it happens: Baby teeth erupt over the course of several months of a baby's early life, while at the same time, the child's adult teeth are *also* slowly growing above and below those baby teeth, beneath the gums. As time passes, the roots of the baby teeth begin to dissolve in a child's jaw, which causes the top of the tooth (the part above the gum) to fall out because there's nothing holding it in the child's mouth anymore. The process is controlled by genetics and might be hurried along as the adult (permanent) tooth pushes from beneath while it's growing into the same space where the baby tooth used to be.

If you've been counting, you might have noticed a mathematical conundrum: how do 32 teeth fit where there were once just 20 teeth? A child's jaw grows faster than does the rest of the skull, making room for the extras, starting with the molars way in the back.

And aside from dentures and implants, those 32 secondary teeth are all you're going to get. While the enamel of a tooth is one of the strongest parts of your body, human teeth are the only parts that can't repair themselves. Compare that with sharks, who get up to 40 sets of teeth in their lifetimes.

The average person spends slightly over a minute brushing their teeth. Assuming that you won't need dentures, if you live a full, average life today (about 85 years old), you will have spent 39 cumulative days brushing your teeth. Still, dentists say that that amount of brushing is less than half of what's needed for a healthy mouth. To that end, just over a quarter of Americans have tooth decay.

William Thomas Green Morton, a dentist in Boston, deserves a big cheesy smile from everyone: in 1846, he proved in public that ether was a useful, safe, and pain-free anesthesia.

Space Science: Planets, Stars, Spaceships

On December 16, 1965, during NASA's Gemini 6A mission, Wally Schirra and Thomas P. Stafford sang "Jingle Bells," accompanied by a harmonica and actual sleigh bells.

Scientists think they know what's on the surface of Venus, but the thick carbon dioxide and sulfuric acid clouds near the planet's surface make it hard to tell for sure.

Here's an argument: There are EIGHT planets that circle the Sun: Mercury is closest, followed by Venus, Earth, Mars, Jupiter, Saturn, Uranus, and Neptune. What used to be called the NINTH planet, Pluto, has been downgraded to a dwarf planet.

At the time of this writing, the Department of Defense's Space Surveillance Network knows of some 27,000 pieces of "space junk" in orbit around our planet; the Federation of American Scientists holds the number in the millions of pieces. Because orbiting waste can travel at extremely high speeds and can crash at those speeds, even microscopic bits of it can be dangerous.

After the Challenger disaster, efforts were made to find as much of the wreckage as possible. To this day, however, not quite half of it has been recovered, and those bits and pieces are buried in old missile silos at the Cape Canaveral Space Force Station in Florida. A roughly 15-by-15-foot (4.6-by-4.6-meter) piece of the Challenger was discovered in late 2022 off Florida's coast, which suggests that what's left likely lies at the bottom of the Atlantic Ocean.

In answer to the age-old question "How's the weather up there?" so-called "blue stars" can reach temperatures of up to 50,000 kelvins, which is nearly 90,000°Fahrenheit. So … um, it's a tad warm.

If you took off your helmet in outer space, you could survive about two minutes with no adverse effects. The thing is, you'd be unconscious after just a few seconds, so you'd better hope you have a friend along to put your helmet back on.

The Kármán line—located a little over 62 miles (100 kilometers) above sea level—is a hypothetical (but legal) line that delineates the point where our planet's atmosphere becomes "outer space." If you got in your car and drove straight upward at about the same speed you'd travel on an average highway, it would take you about an hour to get to space

In 2021, scientists discovered large cyclonic auroras that hovered around the North Pole, spinning and sending electrodes into the upper atmosphere. They called them "space hurricanes." Space hurricanes are similar to other auroras, but are larger and last longer.

One teaspoon (5.69 grams) of neutron star weighs about 10 million tons (a little over 9 million metric tons).

Pack carefully if you're visiting the Moon. Depending on your destination, temperatures will range from 250°F (120°C) near the equator to lower than −200°F (−130°C) on the Moon's poles. Neil Armstrong (1930–2012) said once that the Apollo 11 astronauts didn't go far from their module because of the heat.

Our Sun is a star. Despite its huge size, it's a yellow dwarf.

Should a bit of a star fall to Earth, you could probably pick it up and handle it without being badly burned. If you could somehow reach into the sky to grab a star, though, it would be so hot that it would incinerate you before you could lay a finger on it.

Plants and Animals: Should We Bring Back Extinct Animals? Part 1

Think about this: You want to see a woolly mammoth, but you don't want just bones. You want to see a living, breathing furry animal, but not behind a fence in a zoo or museum. You want to, in fact, populate the world with creatures that haven't seen life in millennia—like, say, that mammoth, a thylacine, a passenger pigeon, or a saber-toothed cat. You can almost imagine that they might someday live again, so how can science make it happen? Is it possible to resurrect an extinct animal, one that's been dead for a thousand years or more?

To get to that answer, we need to split the question: *Can* we? And *should* we?

Let's address the first one: can we bring back extinct animals?

The answer is yes, and no.

The biotech company Colossal is planning to bring back the woolly mammoth by 2027 or 2028, using genetic technology and the womb of a modern-day elephant.

The first thing that has to happen in order to revive an extinct species would be for the creature's genome to be sequenced or DNA retrieved, both of which are easier typed than done.

In many cases, it will depend on how long the animal has been gone. Over time, any genetic matter that can be extracted from the remains of an extinct animal may be degraded to the point where recovery is not possible—which means that extracting the DNA of, say, a stegosaurus is iffy. There are probably many outlying exceptions to this, but most dinosaur DNA has decayed too far for recovery. Also remember: DNA can't be extracted from rocks, which is basically what a fossil is—mineralized remains.

However, if the animal hasn't been gone very long, if the body parts have not been tampered with and are still viable for use in a laboratory, or if someone was prescient enough to have secured DNA back when there were a handful of the creature's brethren alive and well and walking around, then having genetic material to replicate is possible.

But then we have another problem: how do we use what we come up with?

There are three basic ways of de-extinction: cloning, genome editing, and back-breeding.

Cloning is the only way to get a 100 percent *exact* replica of any given creature. When cloning, scientists take a somatic cell, which is any cell in the body that's not reproductive in nature, and they move the DNA from that cell into an ova that is emptied of its nucleus and DNA. The resulting embryo is implanted in a similar animal's womb, and the embryo grows with DNA that is identical to the somatic cell donor. The bad news with cloning is that a total and complete genome is required to perform this process, which may not be possible, depending on when the species went extinct.

Genome editing will get you a creature that's *close* to the extinct species, but not the same. With genome editing, scientists reshape DNA of a biologically similar animal until it mimics the DNA of the creature they hope to bring back. For instance, they might remove the bits of DNA in a domestic cow that cause it to have short hair, and replace it with the DNA of a shaggy auroch, then proceed with the implantation of an embryo, probably in said domestic

FAST FACT

Superintendent of the National Zoologic Park William T. Hornaday (1854–1937) wrote a lengthy essay in 1889 on the extinction of the bison in America. He laid most of the blame on humans, but Hornaday also said that the creature was nearly extinct, in part, because of "the phenomenal stupidity of the animals themselves...." Of course, this was incorrect. The bison was flourishing across North America, with about 30 million animals before the arrival of Christopher Columbus. Today, thanks to conservation efforts, there are about 15,000 bison in America.

cow. The resulting calf would be *a hybrid* of cow and auroch, but it would not be an auroch. To bring back the species would mean many generations of breeding, with a diminishing amount of domestic cow DNA in each generation's genetic material. It would forever be a hybrid; close, even to an infinitesimal degree, but not perfect. It should be noted that scientists have already started trials using genome editing with humans to help prevent diseases.

Back-breeding is a method that takes serious time. It requires the breeding of two different strains of a creature resembling the extinct one, with the aim of offspring that, over time, increasingly resemble the animal that is extinct. Take the auroch, for instance: the magnificent animal went extinct nearly 400 years ago, but some domestic cattle in Europe still harbor auroch DNA. Back-breeding would mean breeding these cattle together exclusively (but carefully, to avoid inbreeding) until resulting calves look and act like aurochs. Again, it's not perfect—the resulting creature is not 100 percent auroch with that smidge of domestic cattle DNA in the old-new breed—but it's something doable by a savvy farmer with mad organizational skills, and it doesn't necessarily require a laboratory.

So we *can* reanimate an extinct animal—kind of, sort of, a little bit, at least almost, and technology is always improving—but the biggest issue is this: if there's any missing DNA or anything that accidentally isn't transferred to an embryo, then what makes the animal *that animal* isn't there, and the whole thing is a wash. A T-Rex without his ferocious teeth, for example, is just another great big lizard.

Earth Science: What If You Fell into a Volcano?

Falling into an active volcano that was about to erupt would be, you can imagine, no trip to the spa.

That boring life you've been living so far would take a sudden turn for the worse if you fell into an active volcano. First of all, the intense heat as you approached the top of the volcano would be nearly unbearable. Because lava is up to 2,200°F (1,200°C) or more on the surface, which is way more than what's needed to incinerate a human body, you'd be wishing for a do-over. After you've fallen in, and on your way down into the opening of the vent (the place from where the lava originates when it spews), your lungs would be burned up so quickly that you wouldn't have much time to scream. The water in your body would start to evaporate, thus causing you to melt from the inside out, and that's all before you hit the molten lava. Once you reached the lava and were burned to ash, you'd be dead. Not fun.

Space Science: Can You Travel in a Straight Line Forever?

Gas up the car, you're going on a trip. If you travel in a straight line forever on Earth, you'll have a few obstacles in your way and a few things you'll have to cross or transverse, but you should end up at the same place at some time sooner or later, ready to do it again.

So, if you traveled on a rocket in a straight line, would the same thing happen?

While we tend to think of the universe as a vast, enormous, unimaginably gigantic void, evidence suggests that the area covered by our universe has limits—big ones, 13.8-billion-light-year ones, but limits nonetheless. Astronomers are still studying it, but it's not out of the realm of possibility that the universe could be like a planet, where you travel in two ways (north or south, east or west) but not in several directions—meaning that one direction in a straight line could return you, sure as a boomerang. It's also possible that the universe is curved, which would make your journey a much longer one (take extra gas) but with the same result you'd get on Earth.

Technology: Shooting a Gun into the Air

You fire a gun straight into the air? Kids, don't do this at home. It might sound impressive, but the truth is that even the most high-powered gun with the most expensive bullets won't get past the Earth's atmosphere, so you're wasting your time shooting at the Moon. You'll never reach it with a mere bullet.

But okay, let's say you shoot your high-powered rifle at a cloud. One bullet, bang. Once that bullet reaches the highest height it's going to reach (possibly as high as 10,000 feet, or 3,050 meters), air and gravity will slow it down. When it has stopped whatever natural ascent it's going to attain, it'll begin falling back down to Earth. Turns out the song is right: what goes up must come down.

While the bullet goes upward faster than the speed of sound, air resistance won't allow its descent to match its upward speed. It'll descend at about 10 percent of the speed at which you shot it into the sky, or roughly 150 miles (240 kilometers) an hour. It'll come down between 30 to 180 seconds after you pull the trigger.

The biggest problem with this is that although you shoot straight up, weather, wind speed, and the

It might look safe, but shooting a bullet into the air is anything but safe! You have no control over where the bullet might land, and sometimes they land on innocent people.

energy and spin of the bullet will keep it from falling directly down to the same place it left the gun. In fact, the bullet could fall on a spot anywhere up to a couple miles away from where you stand, and you can't control this fact.

This is the bad news: depending on what kind of ammunition you use and the kind of gun, that errant bullet you shoot at a cloud could absolutely break the skin of any human or animal that's unfortunate enough to be below it. Studies show that nearly 5 percent of all gun deaths are caused by people who shoot guns into the air in celebration of some event and regret it. Or they don't know they should regret it, and that's too sad to contemplate.

This is why "celebratory gunfire" is illegal everywhere in the United States.

Human Body: A Glutton for Punishment

You take the "all you can eat" part of an all-you-can-eat buffet to heart?

It sounds a little like a dream: all your favorite foods in one place, and you can have as much of them as you want, one price, no limits. So what happens if you plunk down that $12.95, have a seat, and dig in?

You'll be one happy little diner, at first.

Though it's probably very tempting to go for the pie and cake first, most people at a buffet dig into the meat before anything else—perhaps intuiting that sugar will dull the appetite. Meats are often at the top of the list, although salads are right in the middle of several meat choices in a buffet popularity contest. Potatoes in their various forms are almost mandatory for most Americans at a buffet, but rice and pasta won't be missing either. Oh, and on most American buffets, pizza is a must-have.

FAST FACT

Year after year, Joey Chestnut (1983–) enters the Nathan's Famous Hot Dog eating competition and wins it by eating dozens of wieners and buns in 10 minutes' time. In 2019, Chicago Bears linebacker William "Refrigerator" Perry (1962–) went up against Chestnut and lost. Chestnut ate 71 hot dog + bun combinations in the allotted period; Perry had four.

While the length of any huge mealtime can vary, most physicians say you should take your time to eat your meal, at least a half-hour, which is best for your digestive system. But let's say you're mad hungry, you're determined to get your money's worth, you've sailed past that 30 minutes and you're still chowing down....

One of the first effects of eating more than you're used to is that your stomach will expand to accommodate the extra food. This expansion could crowd the rest of your organs, which could be mighty uncomfortable, enough to make you need to unbutton your pants to ease the pain. It's not just your waist that sweats this massive intake, though: the extra food in your stomach is likely going to make your digestive organs work harder, and that's going to include a boost of hydrochloric acid in your stomach. That could cause acid reflux or heartburn. It could also give you a great big case of gas, and because the average stomach can handle about 1.5 quarts (about 1.4 liters) of food, eating nonstop for hours will almost certainly, eventually, trigger your vomit reflex.

All this eating could make you dizzy too, because your metabolism speeds up to help take care of what you're putting in your stomach, and that's going to mess with your blood sugar level. In the long run, if you do this a lot, a big fat meal can make you fat, it'll mess with your sleep cycle, and it could cause other, more serious diseases.

Time to put the fork down: while it's extremely rare, medical science has recorded a few instances of people who literally ate until they died from too much food.

You don't want that to happen, so peep this: they say that the best exercise is to push yourself back from the table.

Human Life: Are You Related to Genghis Khan?

Sometimes, to get to science, you have to open a door to history first.

In the early thirteenth century, the various tribes in southeast Asia were scattered near and far. Still, the Mongols kept track of who was related to whom, alliances, favors, and enemies made. One of the Mongols, Temüjin (1162–1227), was born into nobility in the latter part of the 1100s and he later became the chief of one of the more dominant tribes.

It wasn't without struggle, however. When he was nine, his father moved him elsewhere, having decided that the boy was ready to be married; until he was 12, Temüjin labored in his future wife's family's house. Not long after the wedding, his father was the victim of poisoning by rival Tatars; when Temüjin went to his home village and tried to lay claim to his father's position as chief, villagers denied him the mantle and sent the family packing.

Surviving hand-to-mouth, Temüjin, his siblings, and his mother lived away from their village, but Temüjin's mother put the time to good use. She taught Temüjin about his heritage, politics, strategy, and war. After the death of one of his brothers (at the hand of Temüjin and another brother), capture by former friends of his father's, and subsequent escape, Temüjin joined forces with other warriors and siblings, united the scattered tribes in the area, and became *Genghis Khan*, meaning, basically, head honcho of the Mongols. Over his years as Khan, Temüjin conquered many lands and grew his empire through plunder and war. It was not a pleasant process; many scribes of the time describe the horror of a raid led by Genghis Khan.

And yet, even the fierce must fall: various reports claimed that he died hunting or in battle, but recent analysis suggests that he died of bubonic plague.

Which brings us around to a fascinating scientific question: are you a great-great-great-many-times-great descendant of Genghis Khan?

The answer requires more history.

Like many Mongol men, Temüjin had several wives and concubines—many of them having once been royalty in other cultures, trade merchandise,

A 15th-century manuscript illustration depicts Temüjin being proclaimed Genghis Khan. The title means "master of the ocean," implying that he ruled all lands leading to the oceans—in other words, he was master of the world.

conquered booty, or peacemaking gifts from tribes or villages to the new Khan. It's said that Temüjin spent time with a wife for a night or two before moving on to another camp or tent and another woman. Officially he had a mere handful of children, but DNA suggests quite another story: it turns out that Genghis Khan had children pretty much everywhere he went—one estimate is 500, but it's likely higher because he went far.

Geneticists say that some 16 million men—about 0.2 percent of the world's population—carry genetic material that can be traced back to Genghis Khan. If your family comes from anywhere in what was once the Mongol empire, guess what? Your chances rise to more than 1 in 12 that you are descended from the great and fierce ruler.

Notable Names: Lamarr Was a Star (and Her Genius Went Far)

Proof that you can be wildly famous and secretly wild: film star Hedy Lamarr (1914–2000).

She was born Hedwig Eva Maria Kiesler in Austria, the only child of a father who was a banker and a mother who was a trained pianist. The film industry was in its infancy when she was small, but it's said that she was fascinated by it early on; part of the reason may have been that her father, too, was captivated by the mechanics of it all.

Little girls then, alas, weren't "supposed to" be interested in things like that, and young Hedy was instead schooled in piano-playing, dance, and languages. At age 16 she enrolled in acting school, and she made her screen debut in 1930's *Money on the Streets*. Two years later, she appeared in the movie *Ecstasy*, which made her famous—or, more to the point, *in*famous, since the film included the briefest of nudity and was about as racy as racy got in public in the 1930s. Now the movie is considered to be arty, but when it was released it caused gasps around America, where it was banned.

Married at 18 to Friedrich Mandl (1900–1977), who was much older, Hedy wasn't seen on-screen much for awhile; she supposedly claimed later that her husband didn't want her to act anymore, and it's been said that he tried to find and destroy as many copies of *Ecstasy* as he could.

This control extended to their marriage, with unintended benefits for Hedy: he was a munitions salesman with ties to Nazi Germany, and he socialized with many inside the Italian government, often taking Hedy, who was a math whiz, with him to conferences and meetings he had with various professionals and scientists. At those meetings, she paid close attention and quietly learned about new technologies that various world militaries were developing, which jump-started her interest in technology and applied science. In her spare time, she mentally and very covertly toyed with ideas that came to her, becoming an inventor in various categories, none of which ultimately came to fruition. She tinkered and learned.

A publicity shot shows Hedy Lamarr in 1944 when she starred in the MGM film The Heavenly Body.

FAST FACT

We can't read about Hedy Lamarr's inventing genius without knowing that Gloria Swanson (1899–1983) was also an inventor who held a pile of patents. The star of *Sunset Boulevard* invented a wireless device to summon her butler, for which she held a patent. She was also passionate about working with plastics, metals, and luminous paints; in fact, in the earliest days of World War II Swanson worked tirelessly to bring four German men to the United States to work for her company, Multiprises. Plastics expert Leopold Karniol, metallurgist Anton Kratky, acoustical expert Leopold Neumann, and electrician Richard Kobler came stateside to work on other inventions for Swanson's company, surely saving them from the Nazis.

Still, when she escaped from her husband and fled to America in 1937, Hedy put her scientific interests aside to concentrate on filmdom. She divorced Mandl, and soon after she met the great Louis B. Mayer (1884–1957) and signed a contract to work with him in film. It was he who changed her surname to put room between her *Ecstasy* reputation and her shiny new career. Under his tutelage, she began appearing in movie after movie, becoming simultaneously known as one of the greatest actresses to ever come out of Hollywood and "the most beautiful woman in the world."

Hollywood was good to Hedy Lamarr. She met and briefly dated Howard Hughes, but it's said that she was more interested in his inventions than she was in him as a man. She was supposedly quite inspired by Hughes, who called her "a genius." The most influential person she met arrived in her life at a star-filled dinner party, where she met a man who would put Lamarr on a secret path that didn't become public for many years.

George Antheil (1900–1959) was a musical composer and author who put forth the idea that a woman's pituitary gland could somehow enhance her bust size. That may sound wacky today, but Lamarr was intrigued enough to consult with Antheil, and one conversation led to another. Legend has it that the two began discussing torpedoes and how their trajectory could be vastly improved. That's not exactly dinner-party fare, but keep in mind that the United States was skirting around the edge of World War II. The war was a hot topic.

Lamarr knew about torpedoes and how an enemy ship could jam the frequency of the weapon to throw it off course, making it ineffectual. Antheil, who'd worked for the government as a munitions inspector, also knew about that particular technology, and together they discussed the idea of a weapon that used random frequencies that couldn't be discovered and jammed by enemy submarines.

Eagerly, they began talking seriously. Antheil knew someone at CalTech, Lamarr hired that man, and, collaboratively, the ultimate device was created. In 1942, Lamarr and Antheil received a patent for their weapon, which *they donated,* free and clear, to the U.S. military that same year, not long before the United States entered the war.

And then, very quietly, Lamarr went back to her movie career, selling millions of dollars in war bonds by promising kisses in exchange for money.

Antheil died of a heart attack in 1959. Lamarr's star faded, and she became a recluse in Florida before she died in 2000. Both were inducted in the Inventor's Hall of Fame in 2014, and the Electronic Frontier Foundation named them winners of the Pioneer Award in 1998.

Hedy Lamarr also messed around with ideas for a new kind of traffic light and a new kind of soft drink using a carbonated tablet in water.

As for their invention, it languished for a while because

it wasn't an entirely new technology, and it didn't work well in battle. The U.S. Navy didn't adapt any part of the pair's idea until well after the war was over. Today, as the foundation for the creations of WiFi and cell phones, it's looked upon as a marvel, an idea that was years before its time.

Human Body: Is It Cold in Here, or Is It Me?

So, let's say the temperature gets down to a certain degree. That's about the time you reach for a sweater or a nice warm blanket, unless you want to see how far your body can take you. Then you look at cryogenics.

Most people, when they hear the word "cryogenics," think of bodies suspended in super-cold animation, ready to come alive when the technology ripens for it. But it's not that easy, and dead bodies and man-on-ice are not the only things this scientific field encompasses.

Let's break it down.

Basically speaking, cryogenics is the practical study of and use of materials at extremely low temperatures. We're not talking about a freezer. We're talking about what freezes a freezer: the appliance that keeps your TV dinners solid runs at around 0°F (–18°C); the temperatures that cryogenicists use begin at –112°F (–80°C) and may drop to –220°F (–140°C).

Consider this: any temperature below 31°F (0.5°C) can give you frostbite.

It might surprise you to know that cryogenics is a very old science.

At the end of the 1890s, scientists figured out how to make the elements of air (oxygen, hydrogen,

A metallic tank (called a dewar) is filled with liquid nitrogen at the University of Sydney in New South Wales, Australia. Liquid nitrogen is useful, for example, to store tissues and genetic samples for medical studies.

argon, and nitrogen) into liquid. By 1908, they'd discovered how to do it with helium, which was when they also learned that liquid helium is the coldest liquid ever, at less than –450°F (–268°C). Makes you cold just thinking about it, doesn't it?

Since then, scientists who work in this field have expanded the use of cryogenics into many different areas of our everyday life. The creepy fog you see on movie screens came from cryogenics. NASA uses cryogenics in various space programs on rockets, and the field of superconductivity is a major user of cryogenics. As for your health, MRI machines use cryogenics to get a look at your innards, laboratories use it to freeze blood and tissue samples, cryogenic scalpels can be found in many surgical suites, and cryotherapy (a cryogenic process) is used to freeze abnormal tissue such as skin cancer in order to remove it.

It should be mentioned that a different kind of cryotherapy is gaining a hold on healthcare; that kind of cryotherapy promotes a *very* short time (up to 4 minutes) in a super-cold chamber to enhance

FAST FACT

Your core temperature is generally 98.6°F (37°C), although variations are not unheard-of. That temp is optimal for your organs' survival. If your core temperature drops below 95°F (35°C), you will begin to experience hypothermia. You don't want that.

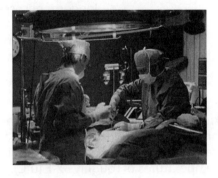

Technicians prepare a body for cyonic preservation in this 1985 photo.

health, but it has its critics. People who use cryotherapy for healing say that it can help sufferers of a variety of ailments because it supposedly alleviates swelling and promotes better blood flow to an injury. The practice, however, isn't proven; it's not a medical procedure, and it's not (at the time of this writing) approved by the FDA.

It's not all fun and games, though: the misuse of cryogenic material can cause asphyxiation, toxicity from the chemicals, and that deep, bone-crushing, tissue-killing cold.

Having said that, there are a couple of possible reasons why you might want to immerse yourself in a freezing space: it's called cryonics, and it's one of the side-processes of cryogenics. The short definition of cryonics is the freezing of entire human bodies (or, sometimes, just an intact human head) with the hope of reversing death at a later date, when whatever caused the death is conquered by a cure. But remember, at this time it's just that: *hope*.

Once and for all, *no*, Walt Disney was not cryonically frozen. He's resting in peace in Forest Lawn Cemetery near Glendale, California.

To offer that hope, and because cryonics is not yet a medically approved procedure, a patient who wants to see what the future will be like must first prearrange this procedure with a cryonics center they can trust to hold their head or entire body in suspension for decades or longer, if necessary.

When they are near death, a cryonics team is alerted so they can be nearby when the inevitable happens.

The second that official death is pronounced, a medical team moves in to restore the circulation and respiration of the client's body artificially and administers substances that will quickly cool it to temperatures of roughly 32°F (0°C). The body's blood is removed, and the water in it is replaced with chemicals meant to keep ice from forming. Ultimately, the body will be kept in a sealed container in long-term storage at temperatures of around −184°F (−120°C) until a time that science and the cryonics firm can rewarm, repair, replace, and reanimate.

It ain't cheap to do this: prices for a whole body are generally in the six figures, with a lower cost in the upper five-figures for freezing just a head (with the *hope* of attaching it to another, probably healthier and younger body at a later time). That may be all-inclusive, with no annual fee. You can find lower prices, and you can find higher prices.

If this sounds like something you'd love to do, keep in mind that scientists freely admit that a lot can go wrong and that they don't know yet how to revive a body that's been frozen for decades or, possibly by then, centuries. The human body is not like a chicken leg, and they don't know if long-term icing might make bigger problems. For sure, delicate cells will probably be too far gone for reuse; early attempts with mammal parts in laboratory settings sound like the stuff of horror novels.

And yet, there's that thing called *hope.*…

Baseball legend and war veteran Ted Williams was cryonically frozen. At the time of this writing, his head and body are in separate, long-term containers at an Alcor facility in Arizona.

Plants and Animals: Ocean Creatures You Should Fear More Than a Shark

Ask any marine scientist, and they'll tell you that most sharks really would rather not have to deal with humans at all. They really don't like to bite us, but they do, often because we move like and appear to be a shark's normal food. Sharks have personalities, they say, but they're not the smartest animals in the ocean.

They're not the most dangerous ones in the oceans either.

 Sharks absolutely have bad press agents. When one accidentally bites a swimmer it's all over the news in a bad way, but the truth is that jellyfish kill many times more

Jellyfish look soft and pretty and harmless, but they actually kill more people every year than sharks do.

people than do sharks. You can, in fact, swim with sharks sometimes without being bothered at all, but jellyfish have translucent tentacles and are harder to see in the water. While it's true that jellyfish don't usually attack people on purpose, brushing up against one that you didn't notice could be really bad: the sting of a box jelly-fish can kill you before you finish your third round of the "Baby Shark" song.

The flower urchin looks like it belongs in a bride's hand on her wedding day, but don't let that fool you. Each of what looks like flowers is, in reality, a tiny little tentacle that can inject venom into whatever part of you happens to get too close. Sharks don't inject venom. Just sayin'.

The wolf eel looks like a really bad nightmare come to life. It's as if a Hollywood special effects master went wild. The wolf eel will grow to longer than 7 feet (2 meters) and to say that they have teeth is a severe understatement; the creature dines on hard-shelled mollusks such as crabs, and that's all you need to know. The good news is that they are curious but not very often aggressive. The bad news is that when they *are* aggressive … those teeth!

The blue-ringed octopus looks as harmless as a guppy. It looks a little like a toy, in fact; it's colorful, pretty, and tiny enough to hide in your hand … that is, if you've got a death wish. The blue-ringed octopus is perfectly happy to bite you and 25 of your friends, and that little nibble is enough to kill you all in a matter of minutes. A shark, on the other hand, is a one-person kind of guy.

One thing about sharks: the predatory ones usually don't hide. Not so with a stonefish, who will lie in wait on the sand beneath the water. A stonefish might even use sand as camouflage, so if you're strolling in a little bit of water on a moonlit night, you won't notice that you've stepped on him until his venomous dorsal fins have already nailed you. The venom could paralyze

Most scientists prefer to call the animal a "white shark" because there's only one species of white shark, nothing "greater" or "lesser" about it.

The stonefish is a well-camouflaged predator that blends into the ocean bottom, waiting for its prey to swim by.

you; if you escape that nasty effect, you may endure excruciating pain that could linger for a long time or even kill you.

The majority of shark attacks happen in water that's over 6 feet (2 meters) deep. The cone snail doesn't need much water at all before you'll be sorry. Just reach for the pretty shell of one of those creatures in shallow water, and the cone snail could reach out with its proboscis and inject its venom, causing paralysis, severe pain, and even death. You can collect shells on the beach, but watch what you touch, and never pick up a live cone snail, no matter how pretty it is. Another thing: not all sharks are dangerous, but all 600 species of cone snails are venomous.

A shark can't bite you if it's dead. But a Portuguese man o' war that died and washed up on shore *weeks ago* still has enough venom to sting you and cause painful welts.

FAST FACT

Brits George (1885–1956) and Ernest (1881–1966) Williamson used their father's invention, a device that could film underwater, to catch a scene in which Ernest fought a shark near the Bahamas. They took the clip to Universal Pictures, which hired the brothers to make the silent film classic *20,000 Leagues Under the Sea* in 1916.

So, then, is it best just to dip your toes in the ocean and stick close to the beach? Maybe: a shark bite is one thing, but a creature with a bite *worse* than a shark's? Ugh, say "no thanks" to a saltwater crocodile, the largest crocodile species on the planet, who will bite you in the ocean, then follow your bleeding body onto land and bite some more....

Plants and Animals: Menagerie, Part 3

Sloths and polar bears living in warmer climates can appear to be green because their fur traps algae. In the case of the sloth, the algae appear to be a form of camouflage from predators that are intent on having sloth for dinner. As for polar bears, the algae is temporary and grows inside the animal's hollow hairs.

Recent research suggests that, before 416,000 years ago, Greenland was lush and warm, filled with spruce trees, lakes, mammoths, and insects. Today, the average temperature of Greenland hovers around 0°F (–18°C).

An elephant's trunk has about 40,000 muscles. Pathetic human, your entire body has just 600 muscles.

At the time of this writing, biologists say that there are TWELVE different main species of foxes, including the swift fox, the fennec fox, the Arctic fox, and the common red fox.

The elephant trunk is truly miraculous. Elephants can lift large tree logs with it, but they can also pick up a pea with the tip, using the 40,000 muscles contained within their nose!

All insects have SIX legs. There are no exceptions.

Elephants have the longest gestation of any animal on the planet. An elephant can count on being pregnant for anywhere between 18 and 22 months. They don't have the biggest baby at birth, though; that honor goes to the blue whale, whose newborn weighs just over a ton when it's born.

Because of the emerald ash borer, a tiny insect whose larvae can kill full-grown trees by cutting off the tree's nutrients, a beloved American sport was changed forever. From the beginning, pro baseball was played with bats made from ash trees. Fewer ash trees mean fewer ash bats, and by spring training season of 2020, most pro baseball players used other types of wood.

A bombardier beetle's name is very appropriate: when the beetle is disturbed or threatened, it shoots a mix of chemicals from an opening at the back end of its abdomen at speeds of up to 22 miles (35 kilometers) per hour. The mixture is hot—up to 212°F (100°C)—and can leave your skin with blisters and bad burns.

Because of their super-slow metabolism, sloths poop about once a week. You can't call them messy, though: they usually "go" in roughly the same place each time.

A squirrel's teeth will grow up to 6 inches (15 centimeters) a year. Gnawing through hard nut shells and seeds keeps those chompers from getting out of hand and growing too long, which could keep the animal from eating.

Those in the know say that you should stay away from wild otters. Apparently, the adorable little guys have a not-so-cute bite.

If the terrain is right, you can hear a lion's roar from 5 miles (8 kilometers) away. Its sound reaches 114 decibels, which is roughly the same loudness as an average rock concert.

Only the outer rings and branches of a tree are alive. The inner rings—the core—are dead but serve to support the rest of the tree so it doesn't fall over.

Notable Names: Judith Love Cohen

Sometimes, in science and in life, you are handed a surprise. Judith Love Cohen was born in Brooklyn, New York, in 1933 and showed an aptitude for math early in her life. By late elementary school, classmates were paying her to do their math homework; later in her education, she was often the only girl in a sea of male mathematics scholars because, well, Cohen loved math. She originally hoped to grow up to be an astronomer, but she didn't think that job was open to girls, and then she fell in love with engineering. While studying at an engineering school in Brooklyn, she also danced ballet at the Metropolitan Opera Ballet in New York City.

After receiving a degree from Brooklyn College, Cohen had two true loves: her new husband and engineering. When the newlyweds moved to California, Cohen reached for more education at the University of Southern California (USC). She attended college in the evening; her days were spent working at North American Aviation as a junior engineer. She also began raising a family and had three children.

In the years surrounding the Cold War and the Space Race, fewer than 1 percent of America's engineers were women. Judith Love Cohen was one of them and she threw herself into her job with fervor, ultimately landing a position with a contractor for NASA. In those years, the projects she worked in included the Minuteman missile program and an abort system for the Apollo Space Program. It was the latter that reportedly made her most proud. For sure, it was the one that made the biggest splash: when the Apollo 13 astronauts lost power on their way back to Earth in April 1970, it was Cohen's work on the Abort Guidance System that got them back to Earth safely.

In 1982, she graduated from UCLA's Executive Engineering Program. Before retiring in the late 1990s, she rounded out her career by working as a systems engineer for the Hubble Space Telescope—

though "retire" is perhaps not the right word. After she left the Hubble project, she launched another career by writing and publishing STEM books for girls, to pique their interest and encourage them to aspire to engineering and science jobs.

In the midst of this storied career, Cohen's personal life changed. She divorced her first husband in the mid-1960s, married again, and resumed having a family 1969. Her oldest son said that she worked on the day she gave birth to her youngest child, taking essential paperwork and her computer to the hospital with her.

You might know that youngest child as movie and TV star Jack Black (1969–).

Former astronaut Sunita Lyn Williams (captain, U.S. Navy, retired; 1965–) once held the record for most spacewalks by a woman and the most spacewalk time for a woman. Peggy Whitson (1960–) holds both those records at the time of this writing.

Human Body: The Five-Second Rule

Ooooops, you dropped your sandwich on the kitchen floor. No problem, though: as long as you scoop it up quick, before five seconds have passed, it's still okay to eat it, right?

Not so fast.

The five-second rule—also known by various other times between three and 15 seconds—is a myth that may have older origins but didn't gain hold in popular culture until the mid-1990s. It states that if food falls on the floor and is picked up within a certain super-quick time, it's still safe to eat.

While it is true that fewer bacteria will glom onto food that gets picked up immediately, *any* contact with a floor—whether it's filthy or has recently been swabbed cleaner than clean—is going to result in bacteria sticking to that food, and it will happen *instantaneously* upon contact. Experiments were done with food dropped on a sterilized, slightly contaminated floor, which showed that five seconds allows plenty of time enough for a lethal dose of *E. coli* to hitch a ride on whatever hit the tile. Imagine, if you will, what carpet, vinyl, wood, or your

The five-second rule that if you pick up dropped food fast enough it will be fine to eat is not strictly true. Bacteria and other contaminants can attach to food upon contact.

FAST FACT

Research suggests that you should remove your shoes before coming into your home because whatever bacteria you trod upon outside will be carried in on the bottom of your footwear and almost instantly get on or into your flooring or carpet. This remove-your-shoes mandate goes double if you have pets or small children, since they're closer to the floor and spend more time rolling around on it than you generally do.

unwashed kitchen linoleum can harbor. Think about where you walked and what was on the bottom of your shoes when you came into your home.

The number of bacteria is going to be affected by what's on the floor where the foodstuff was dropped, how much of the food hit the floor, and the moisture on both food and floor. "Wet" food, such as buttered bread, an apple slice, or a piece of meat, will collect more bacteria than will a potato chip, say, or uncooked spaghetti. Those latter won't be quite as germy when you call "five-second-rule!"

But really ... why take the chance?

Your susceptibility to foodborne bacteria will depend on how your health is. Still, it's true that some doctors admit that they pick up and eat food that's fallen on the floor. Experts say that you should, at the very least, though, wash what you dropped before you put it in your mouth. If you work in a restaurant, sorry, but toss that food out.

Human Life: Hoaxes, Part 1

By now, you've probably figured out one important fact: scientists are pretty smart people. They can also be kind of gullible sometimes.

In 2015, journalist John Bohannan and his colleagues ran a basic (but very legitimate) study with random volunteers, the results of which were "terrible science," just as they expected. They then submitted the iffy information to nearly two dozen journals, claiming that chocolate could boost the effects of a diet. Several of those journals printed unscientific information eagerly and without peer review. The "news" went viral before it was shown to have been a planned hoax.

In 1912, the young world of archaeology was shaken when amateur archaeologist Charles Dawson (1864–1916) asserted that he had made a significant discovery of fossilized remains that "proved" a definite link between apes and humans. Near the village of Piltdown in southern England, he said, in a Pleistocene-era gravel pit, he found bits of a skull that was not *Homo sapiens* but not an ape, either. In the ensuing months, paleontologist Arthur Smith Woodward (1864–1944) joined him in the pits and they discovered more fossilized bones, teeth, and fossilized tools. Smith Woodward hypothesized that the bones were 500,000 years old, and the news was heralded at the next Geological Society meeting. In 1955, long after both Dawson and Smith Woodward were dead, it was proven that Eoanthopus dawsoni, or the Piltdown Man, was 100 percent fake, consisting of a mixture of orangutan and human remains, an elaborate hoax possibly perpetrated by Dawson.

We can laugh now, but two centuries ago, the idea that the Moon was inhabitable was a plausible one. So, along came Richard Adams Locke (1800–1871), who in 1835 submitted a satire to the *New York Sun*, a wondrous story loaded with a scattered few facts that facetiously claimed that evidence of new life forms had been discovered on the Moon. The story—supported by Locke's own "letters to the editor" avowing the veracity of the facts and the "facts"—told of unicorns and flying humanlike creatures, was swallowed lock, stock, and barrel by *Sun* readers. It was picked up by several other newspapers until skepticism won out.

A 1913 sketch of what Piltdown Man (Eoanthopus dawsoni) supposedly would have looked like—if he had actually existed.

FAST FACT

The Cardiff Giant, a 10-foot-tall (3 meter), 3,000-pound (1,360 kilogram) "man," was discovered by workers digging on a project near Cardiff, New York, in 1869. Immediate word was that they'd found a "petrified" man, and the country was all a-tizzy. Scientists weren't fooled, but a good number of the admission-fee-paying public were, as they flocked to see a gypsum-stone carving of a tall humanoid.

Legend has it that Billy the Kid, also known as William Bonney (1859–1881), was killed at age 21 in 1881 by Sheriff Pat Garrett in New Mexico. In 1948, an elderly contemporary of Bonney's claimed that Billy was alive and living as "Brushy Bill" Roberts in nearby Hico. When approached, all Roberts asked for was a pardon for crimes he'd committed; alas, he died before any pardons were granted. The truth is still unknown, but it's said that the folks in Hico, Texas, know and aren't telling.

Human Life: Kissing

No matter what your lips look like, you should be glad to have them. In addition to helping you communicate and helping you attract a mate, your lips help you to chew and swallow, to drink from a can and suck on lollipops, to blow bubbles with your gum and hold a pen in your mouth while you—hang on a sec—are searching for a piece of paper.

And, of course, you can kiss. Ninety percent of the world's human population does.

It seems like a really dumb question to ask, but why do humans press their lips together? The scientific answer is that kissing was once something that mothers did with their infants to determine the health of their babies or to wean them by passing pre-chewed food to an offspring by mouth. Chimps also use kissing to soothe hurt feelings or to deflect anger, and people probably kissed long ago, researchers say, likely as a greeting or as a demonstrative good-bye, so we should take that into consideration.

Those are just evolutionary guesses. As for social kisses, it may be significant that kissing wasn't depicted in art until about 3,500 years ago in India.

That's a wide look over centuries, though. The truth is that some cultures kiss more than others, some kiss less, and others don't kiss anyone, least of all their partners. Though 90 percent of humans in the world do it, fewer than half of the world's cultures find kissing to be pleasant. The rest, which is most of the world, think of kissing as anywhere from "don't do it, thanks" to downright "eeeeuuuwww." You'll find lots of kissers in the Middle East, Europe, and North America. Central Americans and sub-Saharan Africans are those most likely to want to pucker up.

Though it might sound germy (to the average seven-year-old, anyhow), kissing boosts your immunity by exposing you to your partner's bacteria.

Romantic or loving, gentle or passionate, sad or sweet, kissing is a powerful yet puzzling gesture between people.

When you kiss, you transfer billions of bacteria, helping gird you against infection. Kissing releases oxytocin, the "cuddle hormone," plus dopamine and serotonin, all of which bond you to the person you're kissing, thus making you feel connected—which, according to some biologists, might help you choose the best mate for you. Kissing, in fact, might have originally been a way for us to get close enough to sniff one another, to ensure compatibility in mating.

> Despite what you call it when your dog or cat licks you, few animals actually participate in what we know as kissing, meaning the lip-to-lip action. Among those that kiss are chimps and bonobo monkeys, but they kiss for reasons other than romance.

Also, bonus! Kissing burns calories—up to six of them per minute.

So how do you take care of those most exquisite, even erotic, body parts?

Pay attention to the moisture (or lack thereof) in the air. Heating a home tends to make the air dry; dry air is not good for lips. Drink plenty of water, and use moisturizer and sunscreen. Give your lips a weekend rest on the lipstick and tint every now and then. Don't lick your lips, even though it's very tempting to do so. And smile—show off those lips!

Human Life: The Ig Nobel Prize

One of the many hallmarks of good science lies in the proof of concept, and that proof usually comes through experimentation. An experiment can give you the basis for your hypothesis, it can demand another look at the process or theory, and, if you've thought the whole thing through and laughed, it can get you an Ig Nobel Prize.

Every fall for more than three decades, officials at the *Annals of Improbable Research* have announced 10 unique and sometimes silly achievements in scientific research and innovation. The honorees' projects might be serious but trivial, or they might be inventions that make you laugh, or they might be just plain curious and interesting; in any case, the projects feted will "make people LAUGH, then THINK."

The Ig Nobel Prizes were created in 1991 by Marc Abrahams (1956–), the editor of the *Annals of Improbable Research*. The awards were meant initially to poke fun at what might have seemed then to be a handful of ridiculous (but totally serious) big-money-funded projects in categories of public health, engineering, biology, chemistry, and so on. The organization is quick to say that they're not ridiculing science as a whole with these awards; all science has merit to one degree or other, whether it's "good" science or "bad" science. Even so, some Ig Nobels *have* been given out as criticism, and you can well imagine that there's a dose of sarcasm hidden in others bestowed.

For sure, there's humor in the awards, some notoriety, and a little bit of money as a prize, which is a nice bonus. Real-life Nobel Prize winners hand out the Ig Nobel Prizes, which lends the tiniest tone of sophistication to the ceremony. But don't think that this

Marc Abrahams created the Ig Nobel Prizes in 1991.

is all tuxes-and-ballgown stuff: it's tradition to throw paper airplanes at the stage at appropriate times.

Past winners offered studies on malaria-carrying mosquitoes and their choice between Limburger cheese and human feet; a study in which a frog was magnetically levitated; a brassiere that converts to COVID-19–safe face masks; a study on the heart rates of people meeting for a first date; and many other inventions, studies, and ideas that can sometimes become essential parts to a larger (and likely vastly more important) project.

Plants and Animals: Not Just Another Yellow Weed

You grumble about them as you head out to your yard to dig them up. It's a constant battle, and there are days when you wonder why you bother. Technically speaking, what you see in your yard *is* a flower, but you know it as a weed. What's the deal with dandelions, anyhow?

The first thing you need to know is that the dandelion has been around for a whole lot longer than you have: ancient man may have grumbled over it in his yard 30 million years ago, starting in Europe and Asia. Back then, though, humans were just as likely to eat dandelions as anything else; the Romans knew it as both food and an herbal medicine that they could easily grow. Arabian and Chinese physicians grew it as medicine. The earliest Americans brought the seeds with them here because they were familiar with the benefits of dandelions and because the plants were once celebrated for their beauty.

Early Greek and Roman scientists knew the plant as *Dens Leonis,* which morphed into the French *dent de lion,* and the English changed it to "dandelion." In 1753, the plant was classified by Carl Linnaeus (1707–1778) as *Leontodon taraxacum*—the first part being a nod to the plant's leaves, which hark back to its original name and the resemblance to lion's teeth. The plant's updated genus name (*Taraxacum officinale*) reflects its other history: *officinale* indicates the medicinal properties it holds.

There are about 40 species of *Taraxacum* in the world. The common dandelion is the one that's likely most familiar to you and several million home-

Dandelions have many uses. You can use the greens in salads and sandwiches, you can brew teas, and you can even make dandelion wine.

owners as that sunny yellow flower that dots your yard, for better or worse.

You don't have to work hard to get a dandelion to grow. Quite the opposite: you have to work hard *not* to get it to grow. The plant is perennial, meaning that it can appear year-round, though it goes dormant in colder climates. It'll wilt if it gets too dry or too hot, but neither of those conditions will kill the plant. Once it's established, a dandelion plant sends out a taproot that can reach as far as 12 feet (over 3 meters) from the original plant. Be careful how you dig those roots, though: just one inch (2 or 3 centimeters) of root can regenerate a whole new plant.

There is no stem to the plant itself; leaves of the dandelion grow right at the base of the plant, near the soil. In the middle of the leaves, a flower is sent upwards on a single stalk that's 6 inches (15 centimeters) to 2 feet (roughly three-quarters of a meter) in height, depending on the conditions in which it's growing. The flower itself consists of 100 to 300 ray flowers that look like a bunch of skinny petals.

Bees love the nectar inside a dandelion flower.

The most interesting thing about a dandelion happens when the flowers are done. The blossom enfolds upon itself for up to 15 days, and at the end of that period it bursts forth in a rounded, delicate puff of seeds that are carried mostly by the wind as far as 60 miles (nearly 100 kilometers) away. Seeds fall to the ground and replicate rather quickly: today's seeds are next year's crop of dandelions. And if a seed doesn't make it? Well, there's another seed coming along: one single dandelion head can hold up to 200 seeds, each plant can offer up to 20,000 seeds, and dandelion plants can live for several years.

There are many reasons to stop grumbling about dandelions, and even more reasons to let them live, at least for a while.

As mentioned, ecologically speaking, dandelions are good for the bees and insects who visit

FAST FACT

When housing developer William Levitt (1907–1994) began building his famous "Levittown" suburbs, he made sure that each house was surrounded by lush, green growing lawns, no dandelions allowed. Today's Americans, overall, maintain around 20 million acres of lawns.

them for nectar. They're also good for your lawn because the roots loosen and aerate soil and pull deep nutrients closer to the surface of the ground, making those nutrients easier to access for shallower-rooted plants.

Not only can you make beer and wine from dandelion parts, but you can eat dandelion greens in your salad. Pull them when they're young, or they might taste bitter. You can also eat the blooms and use the roots for making tea. Dandelions are full of vitamins A and C, and a whole lot of calcium, iron, and potassium. Just make sure that what you harvest hasn't been treated with pesticides or herbicides.

As for the fabled medicinal properties of dandelions, it's true that the plant helps the liver to filter toxins, can be used as a diuretic, and can assist the digestive system. New studies suggest that the dandelion also has antioxidants and anti-inflammatory properties, and that it can help with blood sugar regulation too.

Not bad, for a yellow weed, eh?

Notable Names: Alexis St. Martin Just Couldn't Stomach That

Sometimes an accident is more than an oops. Sometimes an accident turns out to be a big scientific boon.

Alexis Bidagan St. Martin (1802–1880) was just 20 years old when his life radically changed in an instant. St. Martin was a Canadian voyageur, a man tasked with transporting furs on northern waterways during the peak days of the North American fur trade. There he was, at a trading post on Mackinac Island near what is now Michigan, when it happened—boom!—someone shot him at close range, right in his abdomen.

It was 1822, and almost any kind of gunshot wound was a serious, if not deadly, matter. Indeed, it surely appeared that St. Martin was doomed. How could he survive with a hole in his stomach, mangled abdominal muscles, broken ribs? When the poor man ate, food fell out of his wound before it could be digested. How could he be expected to live?

Nevertheless, U.S. Army surgeon Dr. William Beaumont (1785–1853) was immediately called, and he set about doing what he could. He tended to St. Martin's wounds, bled him (meaning that Beaumont caused St. Martin to bleed some more), and then gave him a cathartic, all of which were standard medical care then. As if this wasn't literally adding insult to injury, becoming injured meant the instant loss of his job, and Mackinac Island officials wanted to send St. Martin back to Quebec.

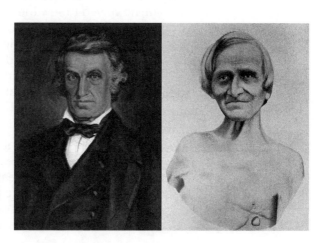

Dr. William Beaumont (left) and his patient, Alexis St. Martin

Beaumont resisted doing so. He honestly didn't think the man would live to finish what would be a long, arduous journey. Instead, he insisted that St. Martin stay, and he continued to minister care to the unfortunate man.

Astoundingly, St. Martin lived that first day and many days afterward; he, in fact, apparently became well enough to eat and drink just days after his injury, which Beaumont reported with great detail. Sometime in the middle of the second week, St. Martin healed enough so that the food he ate began to stay in his digestive system. He still had a large hole in his abdomen, but the wound was looking better, his bowels began to work again, and he was alert.

Beaumont recognized a golden opportunity. He hired the newly itinerant St. Martin to work in his household, and after procuring written permission, he began experimenting on the illiterate man.

Accidentally, the edge of St. Martin's stomach had attached itself to the edge of his skin as it healed, leaving a gastric fistula, or a sort of a tube between inside and outside. Beaumont could literally see St. Martin's body working, and he seized the opportunity to devise further tests to learn about digestion and the digestive system functions. At one point, he vowed to sew the fistula up and let St. Martin live his life, but he never did. Even so, and though a hole in his abdomen surely changed the way he existed, St. Martin eventually married and fathered several children.

William Beaumont got his medical experience through an apprenticeship, as did most of the doctors of his day. He became a surgeon during the War of 1812.

Beaumont published a booklet in 1838, explaining what he had learned, which was radical info then. The knowledge he'd gleaned about digestive acids was especially

SCIENCE

noteworthy; it meshed with other scientists' understanding at the time, proving a lot of theories.

Sometime in the earlier 1830s, St. Martin left Beaumont and returned to Canada, refusing to come back to Beaumont's side, no matter where the doctor went. Beaumont died in St. Louis, Missouri, in 1853. St. Martin, the man who wasn't supposed to live at all, outlived his doctor by more than a quarter of a century.

Human Body: Reproduction Roundup

- *Women knew all* about menopause and Greek physicians acknowledged that it was real, but it wasn't until 1821 that there was an actual word for it. That was the year when Charles-Pierre-Louis de Gardanne, a French doctor, used the words for "month," or moon, and "stop" to create "menopause."

- *Very early in* their gestation, male and female human embryos are identical, except for the chromosomes that will send hormones to start changes in reproductive organs. Also very early in gestation, human embryos and puppy embryos are remarkably alike.

- *A female baby* is born with all the eggs she'll ever need. Those eggs were formed long before she was born, so while your grandmother was pregnant with your mother, your mother's ova were forming—thus, while your grandmother was carrying your mother, she was also carrying you.

- *It is possible*—though extremely rare—for a woman to be pregnant with two different fetuses at the same time. Called superfetation, it happens when pregnancy hormones fail to stop a second ovulation. It's also possible, due to congenital issues, for a woman to have two uteri and get pregnant with both at different times.

- *Studies show that* not long after birth, infants cry in a manner that mimics their mother's native language.

Space Science: Living in Space on Earth

A terrarium is a place for your pet lizard. It's a place to grow awesome orchids. But is it a place to nurture humans?

When billionaire philanthropist Ed Bass (1945–) met ecologist/engineer John Allen (1929–) in the 1970s, it must've been instantly obvious that the two had similar ideals: both were extremely passionate about ecology and Earth sciences. Both men had dedicated themselves to being entrepreneurs; Bass was an oilman and Allen was the founder of the Synergia Ranch (a place for ecology-based innovation and creation) in New Mexico. Both had money to invest.

By the early 1980s, the two men were not only friends but collaborators. Allen's crew at Synergia Ranch had been tinkering with ideas for a self-sustaining way to keep humans alive on other planets, often with funding from Bass. Biosphere 1 was, of course, Earth. With Bass's money and Allen's know-how, they cofounded the Biosphere 2 project that broke ground in 1987 and was completed by 1991.

The original idea for Biosphere 2 sounded lush and lovely, literally like heaven on Earth.

FAST FACT

Spaceship Earth was a phrase that began with Buckminster Fuller to describe the way he perceived our planet. He saw Earth as a giant spaceship and the humans here as one very large crew that needed to work together, as would the crew of a large cruise ship. He believed that it was our responsibility to cooperate to keep the "ship" working properly and that the Earth's resources would one day be equally shared between all its "crew."

Laid over 3.14 acres (0.012 square kilometer) of Arizona land a little north of Tucson, Biosphere 2 was made mostly of a steel frame and more than 6,000 glass panels that were sealed extremely tightly, so that nothing escaped through the walls or roof and nothing got in that didn't belong in. To accommodate the expanding and shrinking of the walls caused by the Sun's heat, Biosphere 2 had what the builders called "lungs" to shrug off the extra volume.

Inside, there were several distinct areas (biomes) characterized by climate and by the creatures that live there—among them were areas simulating a rainforest, a coral reef, a wetland, a savannah, and a fog desert. These biomes were intended to be populated by eight humans who would live there during an experimental period of two years; there was also a large agricultural area for growing food; 3,000 species of plants, insects, and livestock animals; and a laboratory so that experiments and studies could be done within the biodome. Below ground was a heating and cooling system and other mechanics; the Biosphere 2 also relied on passive solar energy through glass panels that covered most of the closed-system area. Electricity was supplied as well as natural gas.

In late September 1991, eight crew members, including a medical doctor, entered Biosphere 2 and were sealed in, to great fanfare. The media was there, as were a lot of tourists who were allowed to walk the grounds outside Biosphere 2. A lot was riding on this experiment, including a few new inventions that their creators hoped would lead to patents. This project came with big hopes. Really big ones.

Within two weeks, there were problems. One of the crew experienced a medical emergency and was removed for surgery. She returned later but not without additional supplies that weren't there at the outset, which happened again later, resulting in minor scandals that ultimately became bigger deals than perhaps they would be today. It should go without saying that the news media

The public is now able to visit Biosphere 2 in Oracle, Arizona, where the facility is used for research and education.

watched Biosphere 2 with eagle eyes, and no small incident escaped notice.

It quickly became apparent that there was a problem with the food supply: agricultural crops meant to provide the majority of their meals didn't grow as well as expected, and the crew reported constant hunger. While most of their diet was vegetarian, provided by what they grew and supplemented by the animals inside Biosphere 2, reports were that emergency food had been stored away—another thing the news media didn't like. The crew reported problems with cockroaches, which had accidentally come inside and which bred like ... well, like cockroaches. Ants were also among the first unintended irritations. Still, from a standpoint of workability and biome thriving, Biosphere 2 was initially a rousing success.

Remember, though—this was an experiment that was ultimately about living in space. Humans in space couldn't exactly leave if they suddenly didn't like their dome-mates, so, as for the human residents, Biosphere 2 was an unintended lesson in psychology.

While there were delights to be had—it must've seemed like a modern-day Garden of Eden—the eight crew members separated into two groups of four almost instantly. There was a power grab, reported leadership problems, and friends became enemies in what psychologists call "irrational antagonism," which is a phenomenon that happens when people are forced to be together in a small space for longer than about six weeks. Reportedly, there were no fisticuffs, but it wasn't always pleasant.

The biggest problem presented itself a few months after the crew entered Biosphere 2: they realized that the oxygen levels inside the domes were falling too far, too fast. Altitude sickness set in for some. Ultimately, oxygen was pumped into one of the biomes to avoid having to abort the mission; later, it was discovered that microbes had produced more CO_2 than the plants could process. Managing CO_2 levels seemed from then on to be a constant chore.

Exactly two years to the day after being sealed in, the first crew left Biosphere 2 in good health, despite the problems. While the experiment had gained huge amounts of scientific knowledge in several different "ologies," most people seemed to think there should have been more.

With the bugs mostly worked out (literally and figuratively), a second crew of seven people entered Biosphere 2 on March 6, 1994, intending to stay just ten months. Almost immediately, issues that had come up during the first experiment roared back to life.

A few days after the new crew were sealed in, financial and management issues rose and then-financier Steve Bannon—who'd been hired by Bass to run Space Biospheres Ventures, the company that owned Biosphere 2—became heavily involved in the project, which was said to have been losing money.

Before the first crew had even exited Biosphere 2, Bannon had decided that a change in management was needed and had fired much of Biosphere 2's original leadership. This led to two of the first team's members breaking into Biosphere 2 and doing a little vandalism to the glass panels in protest. Cue the lawyers, who quickly got involved.

The second stay in Biosphere 2 was aborted a few short months after it began.

Was Biosphere 2 a colossal waste of money, or did it bear scientific fruit? That depends on who you ask, because it was successful in some ways and not in others. It failed, but not totally, and there may yet be relevant things to learn from it.

In the waning days of Bass's ownership of Biosphere 2, Bannon made a deal with Columbia University to run the Biosphere. Some time after 2003, the University of Arizona took over the job, and in 2011 Bass donated the whole enterprise to the University of Arizona and set up an endowment fund. Today, the University uses Biosphere 2 as a laboratory and research center, and visits are allowed every day of the year except Thanksgiving and Christmas.

Human Body: Billy Rubin Is Not a Stage Name

Your body is an exquisite machine that requires maintenance from time to time. It's not always fun even if it sometimes sounds as though your doctors are working on a Broadway stage.

If your doctor says you need a Billy Rubin test, that's not an indication that you should break out in song and dance. Bilirubin (pronounced the same way, see?) is a reddish-orangey-yellow compound that's a natural by-product that your body creates. When a blood cell is in a state of disrepair, is getting old, or was abnormal at the outset, molecules fall off the cell. Breakdown of the blood cell occurs, and bilirubin is the result.

No matter where it comes from, bilirubin ends up in the liver to be processed. From there, it's excreted in urine or feces, depending on where the breakdown occurred. This is all done without your direction or intent. You don't even have to flip a switch.

That is, unless something is amiss. If bilirubin levels are high, that could indicate to your doctor that your liver or bile ducts are having trouble. High bilirubin rates mean that more blood cells are being destroyed and processed, which can cause jaundice, or the yellowing of the skin—something that's relatively common in newborns. It could also mean that your liver is struggling to process bilirubin overall; other out-of-the-norm levels of bilirubin can indicate hepatitis, anemia, and other liver diseases.

Kupffer cells are found in the blood vessel walls of the liver and act as a kind of immune cell there. Kupffer cells are named after German anatomist Karl Wilhelm von Kupffer (1829–1902), who noted them in 1876.

So why doesn't science look for ways to get rid of bilirubin altogether?

It's true that bilirubin is a waste product, but your body gets rid of it all by itself without the help of scientists. Also, studies show that bilirubin might be a bit of an antioxidant and in small amounts may be good for your heart and your circulation.

Absolutely, *this* Billy Rubin is a good actor.

Earth Science: Big and Small

Of all the kinds of rocks that exist on Earth, clay is the smallest. Clay particles are just shy of 2 microns in size. For comparison, a single human hair is 75 microns in diameter.

The largest rock (also called a monolith) is Uluru or Ayers Rock, located in Australia. Made of sandstone, Uluru is 1,142 feet (about a third of a kilometer) high, a little over 2 miles (a little over 3 kilometers) long, and 1.5 miles (a little over 2 kilometers) wide. Geologists say that a large portion of Uluru, which was formed some 350 million years ago, can't be seen because it's underground.

Nature's smelliest rock is antozonite, which is a calcium fluorite mineral that, when crushed or scratched, emits a really bad smell. The stench comes from fluorine gas inside the rock, giving antozonite its nickname of stink-spar.

If Charlie Brown got a plain old *rock* for Halloween, chances are that it was a sedimentary rock. Sedimentary rocks are the most common, run-of-the-mill rocks on Earth. Quartz, by the way, is the world's most common mineral.

The International Gem Society lists several gems as being rarer than diamonds, but the story of the gemstone painite is an unusual one. Painite was discovered in 1951 in Myanmar as a *single* crystal and that was it until 2001, when two more single crystals were found. At the time of this writing, a little more than a thousand painite crystals have been located, but few of them are able to be faceted for the sake of jewelry.

Named after British mineralogist Arthur C. D. Pain, painite is an extremely rare borate mineral worth about $50,000 per carat.

The lightest rocks around are reticulites, which is a kind of pumice stone. Pumice is an igneous rock that's so light it can literally float on water.

You'll get a lot of arguments if you try to learn which rock is the heaviest. The reason: "heavy" is relative to size and the amount of metal in the rock, so answers will vary. If you want to know the hardest, though, that would be a diamond.

Head to Canada if you want to see the world's oldest bedrock. It lies on the northeast coast of Hudson Bay, and geologists say that it's 4.28 billion years old.

One of the weirdest rock-like surprises is a fulgurite, which is formed when lightning strikes the Earth at temperatures of 50,000°F (28,000°C), which melts pretty much everything it touches and fuses sand or rocks together. Sand fulgurites occur mostly on beaches and can be more than two dozen feet long; rock fulgurites form in more mountainous areas and show up on the surface of rocks. Fulgurites are found everywhere on Earth, but they're not very common.

*Obsidian—volcanic glass—*is so sharp when fractured that today's surgeons sometimes use it instead of a regular scalpel to perform surgical procedures.

If you're talking rocks, how do you differentiate between them in size? In the 1920s, geoscientist Chester Keeler Wentworth (1891–1969) altered a scale invented by Johan August Udden (1859–1932) and created the Udden-Wentworth grain size scale. Basically, rocks are differentiated

FAST FACT

Karat or carat? A carat is a unit for measuring weight that you use to tell your friends how big the gemstone is that sits on your finger. A karat is how much gold is in an alloy of 24 parts—for instance, 18k gold is 18 parts gold, 6 parts something else (usually zinc, copper, or another metal).

FAST FACT

Scientists say that music—all music, from rock to classical to hip-hop to raga to Tibetan chant—improves mood and helps us with focus, relaxation, and communication. They're still studying why and how but they believe it may have something with the fact that music causes our brains to release endorphins.

thusly: a boulder is a rock fragment larger than 10 inches (25 centimeters) in diameter. "Cobbles" are 2.5 to 10 inches (6 to 25 centimeters) in size. Sand can range between .002 and .07 inch per grain (fractions of a millimeter). Clay particles are the smallest; you need a microscope to see an individual grain of clay.

More specifics: in the early 1800s, German chemist Friedrich Mohs (1773–1839) created a scale to determine the hardness of minerals based on how easy they were to scratch. Talc, he found, was the softest mineral, while the diamond was the hardest. His work today is called the Mohs scale, and its 1 to 10 ratings (softness to hardness) are still used by geologists.

Rocks you never want to collect: arsenic, realgar, cinnabar, and orpiment. All of these rocks contain substances that are dangerous to handle, and you should never get any of them close to your face. They're pretty rocks. And they're pretty deadly.

Human Life: Hoaxes, Part 2: The "Gentle" Tasaday People

Toward the end of the Vietnam War, an exciting discovery was announced that stunned not only the scientific community but the world as a whole. In July 1971 Manuel Elizalde Jr. (1936–1997), head of the Philippine government's PANAMIN (Presidential Assistance on National Minorities), an organization formed to protect cultural minorities in that nation, claimed that he was part of a discovery that was ground-shaking.

According to Elizalde, a small group of individuals—roughly 25 people—had been found in 1966 by people of another nearby tribe, a fact that had been kept quiet. The families were living primitive lives in caves in the jungle near Lake Sebu in Mindanao; Elizalde was glad to realize that they were peaceful, but they seemed to have little knowledge of other humans outside their own small group and the few outside tribesmen they'd interacted with. As for the outside *world*, they had had no contact and thus no idea of what awaited them.

Elizalde further ran with the story, telling reporters that this tribe, the Tasaday, had been living the lives of "cavemen," unaware that there'd been wars, ignorant of the existence of an ocean, and totally in the dark about lights, telephones, and everything else that had been invented while they lived a fishing, foraging "Stone Age" kind of life. He said they had tools that they made themselves, and they lived basically just as our ancestors lived.

Immediately, journalists flocked to the Philippines to report on this amazing discovery. *National Geographic* magazine and the *New York Times* both reported on the Tasaday tribe. Books were written about the Tasadays. Films were made. People couldn't get enough of them.

The first cracks in this great scientific discovery came in the early 1970s when Elizalde seemed to appoint himself as a sort of guardian for the Tasaday. He suggested to Philippine President Ferdinand Marcos (1917–1989) that visits to the tribe be kept at a minimum, which caused a bit of rumbling in the scientific community. In 1974, Marcos shut everything down and that was that—no more visits to the Tasaday.

Well, only for a while, it was.

While the Ta-saday are a real people with a distinct dialect, they were never the primitive tribe as claimed by Manuel Eli-zalde Jr.

When President Marcos was deposed in 1986, the whole Tasaday story fell apart like a cheap toy. Swiss journalist Oswald Iten (1950–) visited the Tasaday once there was no longer a barrier to doing so, and he found the Tasaday living in modern homes and wearing modern clothing. They claimed then that, years before, Elizalde had bribed them to wear crude loincloths, to cook animals over open fire, to live in caves, and to otherwise act like they'd come from the Stone Age. They said that they were promised some level of protection from nearby warring tribes and money in exchange for the roles they'd played.

So the Tasaday-as-primitive-people story was a hoax.

Or it was real.

Which was it?

The truth, according to many anthropologists, is somewhere in the middle: absolutely, yes, the Tasaday are a tribe of people. By their own accounts, they were paid to act like "cave men." They *were* probably isolated at some point, but it's unlikely that they were isolated for a few centuries or so. So there were lies and deception, but they were, indeed, an authentic forest-dwelling, mostly modern-life-shunning tribe.

Human Body: Sexual Anomalies

Sometimes, things happen to make you scratch your head. Few things in life are solid, unwavering, or unchangeable. Wouldn't you know, that includes humans?

The Güevedoces

Back in the 1970s, endocrinologist Dr. Julianne Imperato-McGinley of Cornell Medical College in New York heard about a mystery in a tiny little town in the Dominican Republic. She headed south to investigate and found that the story was true.

In this town, which is not named to protect the privacy of all concerned, many baby boys are born with genitalia that looks somewhat ambivalent but that appears more female-like. Those children, whom the locals call "Güevedoces," are given feminine names, feminine clothing, and dolls to err on the side of caution. Still, many Güevedoces grow up insisting that they are not girls—and they're correct. At puberty, their testes descend and penises grow. The "girls" are boys after all.

At conception, most people have two chromosomes—two X chromosomes for those likely to be girls, or an X and a Y for those likely to be boys. If you looked at a very young (under eight weeks) fetus, you'd see no difference because all early fetuses look the same. It's at that eight-week mark or thereabouts when hormones kick in and begin the fetal body on its transformation to the gender it will be born with.

All was typical with the Güevedoces early in their conception, but Imperato-McGinley discovered that each of them was missing one thing: a full dose of an enzyme called 5-alpha-reductase, which meant their bodies would not convert testosterone properly. The babies looked like girls at birth, but once puberty arrived and the children were suddenly awash in testosterone, they outwardly became the boys they were, genetically, all along.

Studies show that after puberty, most Güevedoces identify as men and live their lives as heterosexual men. A small number of them don't, though, and they're embraced by their community just the same. Treatment for this genetic anomaly exists, but there's no word on how many parents ask for it early on.

Two Uteruses?

Imagine your surprise when you've taken your newborn baby home from the hospital and all is well, but you begin to have mysterious pain, only to find out that you're still pregnant.

It can happen to women who have two uteruses, which is not as rare as you might think it is: didelphys (the scientific name for the phenomenon) affects about .03 percent of the population.

When a female fetus is developing, the Müllerian ducts in her budding reproductive area usually fuse into one, making a uterus. If the tubes don't fuse properly, the result is two uteruses. Some women also have two vaginas and two cervixes.

In many cases, the situation can be fixed. If not, well, infertility might result, or you may find that both uteruses work just fine.

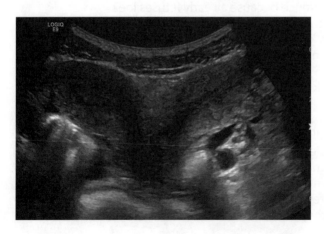

An ultrasound of a woman with didelphys shows she possesses to uteruses.

In 2006, a British mother gave birth to three babies, one from one uterus and identical twins from her other uterus. Doctors called it "risky," but all three babies were healthy after a Caesarean-section birth.

In 2019, a mother in Bangladesh brought home her healthy newborn boy, but almost a month later she was back at the hospital with mysterious abdominal pains. That was when she too gave birth to a set of twins. Doctors said that an ultrasound would have shown them three babies, had the mother had an ultrasound, but she lived in a poor rural area with no access to such medical care.

And speaking of unique pregnancies, in 2022, a Texas woman ovulated again after she was already pregnant, a phenomenon called superfetation. The babies, both boys, were conceived a week apart from separate eggs and sperm but were born on the same day. Superfetation is so exceedingly rare that only a very small handful of superfetation pregnancies have been recorded worldwide.

Joseph Pujol, Professional Farter

If you look at it strictly head-on, the early years of Joseph Pujol's (1857–1945) life were rather unremarkable. Born and raised in France, he was just your typical average boy when, as the story goes, Pujol felt a rush of water into his rectum while he was swimming. Shocked, to say the least, he leaped out of the pool and realized, probably to his embarrassment, that water was pouring out of his bottom. Lots of it.

Okay, that turned out to be a cool parlor trick: as an adult, Pujol loved to entertain his army buddies by sucking up a large amount of liquid in his rectum and shooting it back out, neatly and at an impressively far distance. Then he discovered that he could strategically pull in air and push it back out so that it sounded like a musical instrument.

After leaving his military career, Pujol became a baker—but really, would you let this kind of talent

Stage performer Joseph Pujol made a living with his remarkable ability to maintain absolute control over flatus. He could even play tunes with the methane projections from his behind.

go to waste? Nope, and in 1892, Pujol appeared for the first time at the Moulin Rouge. By this time, he was married and had started a family.

Audiences were delighted to see Pujol, who performed under the name Le Pétomane, as he (odorlessly!) showed them what he could do. His stage act was varied but it featured a round of fart-impressions and the insertion of a tube in Pujol's backside through a hole in the seat of his pants. He'd then ingest and then divest himself of liquid and smoke cigarettes from behind. He let loose with a volley of sound effects, "sang" recognizable songs from his colon, played musical instruments via that rubber tube inserted inside himself, and blew out candles. It's said that audiences laughed until they practically had to be carried out.

Mere fame in France wasn't enough for Pujol, however, so he went on tour all around Europe and North Africa. He performed for everyday folks and for kings and wise men, and he was, for a time, the highest-paid performer in all of Paris. When the tour

FAST FACT

Because they are ruminants that chew their food twice, cattle belch *a lot* of methane into the atmosphere and, for that, are the top agricultural source of greenhouse gases in the world. Billy goats, on the other hand, are also ruminants but have better digestive systems and thus are not very big contributors to the methane problem.

ended and it was time to take his butt home, he decided to open his own act, independent of the Moulin Rouge, with a more refined performance.

But tastes change and so do the tastemakers. At the beginning of World War I, Pujol was said to have been discouraged and disheartened by the war itself. He'd aged by then, and performing didn't hold much for him anymore.

By the end of the war, Le Pétomane had hung up his rubber hose and gone back to baking bread.

Human Body: Waste Not, Want Not

If you are like most people, you eliminate your body's waste in some sort of porcelain receptacle; in other words, you use a bathroom with running water or a compost toilet. But there are some animals that really wish you wouldn't. (Caution: what you're about to read has a big "ick" factor).

Pee

First, let's look at what's in your urine.

When you urinate, you cast off potassium, nitrogen, and phosphorus that your body doesn't need anymore. Nitrogen contains lots of urea, which also contains sodium. The more dehydrated you are, the more concentrated the urea in your pee.

Every animal on Earth needs sodium to live, and carnivores generally get it by consuming the meat and blood of their prey. Generally speaking, a successful carnivore doesn't worry a bit about its sodium intake, but herbivores can never get enough salt. This is why farmers will put out salt licks for their cattle, horses, and sheep: because plant roughage doesn't have enough salt in it to sustain life. Wild animals will do whatever they have to do to get enough salt in their diets. That includes eating dirt,

FAST FACT

John W. Armstrong (b. 1881?) strongly believed in the benefits of natural medicine. Recalling that his family regularly used urine for folk cures, he published a book in 1944 called *The Water of Life*, advocating what he called Urine Therapy. Yes, followers of Armstrong's book are told to drink their own urine, despite there being no proven therapeutic benefit from doing so.

traveling around the countryside to find natural salt, or finding a human who urinates on the ground.

Mountain goats have been known to follow hikers around, waiting for the inevitable. Moths and butterflies are drawn to puddles that aren't necessarily rain runoff. Urine, for the Tozhu people of Siberia, is one way to bond with their reindeer: they purposefully pee in places where their reindeer can easily find it, which also serves to acclimate the animals to the scent of their humans. Locusts love urine that's been sitting around for a few days.

Poop

Now, what's in an animal's fecal matter?

The waste from an animal is going to depend on what it eats, whether it's carnivore, herbivore, or omnivore. It's also going to depend on the natural pathogens, bacteria, and viruses that animals of its species carry, as well as the animal's dietary needs.

On that note, coprophagia is the word scientists use to describe some animals' feces-eating habits. This probably really sounds beyond disgusting to you, but the thing to remember is that consuming such waste can absolutely be a matter of survival.

Lagomorphs, such as rabbits and hares, are rear-gut fermenters, which means that their food is digested faster than the nutrients are absorbed. Because of that, lagomorphs rely on pooped-out cecotropes, which are soft and full of nutrients that the animal's body might have missed the first time around. You may never see cecotropes because they're mostly excreted at night and the animal generally consumes them immediately.

When they are being weaned and are in a sort of transition, piglets, hippopotamus babies, and elephant calves will eat their mothers' feces in order to pick up beneficial bacteria in their gut, bacteria they

Some species such as rabbits will eat their own pellets because the partially digested food still contains nutrients for the animal to absorb.

didn't get when they were born or during nursing. Scientists have evidence that that essential practice has been going on since mammoths were around.

Dogs seem to have a propensity for eating waste, human and their own, and several theories have been put forth to explain why. They'll consume their own waste because of anxiety, for attention, or in response to abusive situations or medical issues such as canine dementia, and stopping the behavior takes determination and patience. As for consuming human waste, wolves don't do it at all, so it's been suggested that the act has something to do with evolution and living with humans.

So, what's in your poop?

Roughly one-third of your solid waste is dead bacteria from your digestive system, and a third of it consists of indigestible material such as corn cellulose, vegetable matter, and so on. Up to 20 percent of the remainder is cholesterol and other fats; another 20 percent is chemical in nature, such as phosphates; and the rest is largely protein, dead cells, pigments, and water.

For many centuries, pigs have been used in "pig toilets," which is just what it sounds like: a feed trough positioned beneath an outhouse. Human waste is not the pig's first choice in food, but a pig—though an otherwise clean and known-to-be-smart animal—will eat basically anything if it's hungry. And pigs are rarely not hungry. Pig toilets, by the way, sound really gross, but they are still in use in many underdeveloped countries and rural areas therein.

And finally, you can't have an entry like this without a happy ending.

The world would be a terrible place without dung beetles.

Found on all continents except Antarctica, dung beetles are the street-sweepers of the animal kingdom; they mostly like to do their work in climates that are temperate, with a little moisture from the sky now and then.

There are three kinds of dung beetles: Rollers will literally roll dung in large spherical balls, which they move with their hind legs to a nesting area, which takes enormous strength; one species can move over 1,100 times its own body weight. Rollers lay their eggs inside the fecal ball, and when the young hatch, they eat the fermented matter. Tun-

FAST FACT

In early 2018, Microsoft inventor Bill Gates (1955–) gave a speech at the Reinvented Toilet Expo in Beijing to discuss new technology in sanitation. He brought along with him a sealed glass jar of human fecal matter, to drive home a point about how much disease and bacteria it contained. The jar sat on a shelf next to the podium for the entirety of the speech.

nelers will bury dung on the spot where they find it and use it for the same purpose. Dwellers just move in, as if the feces were some sort of new condo.

In doing all this work, dung beetles perform several different beneficial things for the planet. They fertilize the soil and help spread seeds. In taking the fecal matter away from the livestock that produced it, dung beetles help keep flies away from livestock, and that helps reduce disease in the herds. And by processing so much waste, the dung beetle removes it from the surface of the Earth, making it harder for humans (or any creature) to step in it. Ugh.

Physics, Chemistry, and Math: This and That and the Other Thing

The body of Marie Curie (1867–1934), a Polish-French physicist who discovered radium and worked extensively with radium and polonium, is still radioactive.

Say there was a brief snow shower overnight, but when the Sun rose first thing in the morning it melted the snow, and the ground steamed a little from it. That's one example of a unique thing about water: it can exist in three different forms at the same time and place (ice, vapor, and liquid).

How many colors are there? Chances are, you grew up with a palette that depended on a crayon box, or you knew ROY G. BIV (a mnemonic for the colors of the rainbow in order: red, orange, yellow, green, blue, indigo, violet). The truth is unknown, but it's believed that there are billions of billions of different colors, most of which are outside our ability to see.

That plastic bag you used to carry your groceries home from the store will totally break down and biodegrade in about two decades. The plastic soda bottle in your fridge will take considerably longer, at more than 400 years until decomposition.

The FAA allows just one small medical thermometer per airplane passenger, and it *must be in a protective case* in a carry-on bag. The reason? Airplanes are made of aluminum. Mercury can very quickly corrode aluminum.

The half-life of DNA is 521 years.

Chemist Luke Howard (1772–1864) was the first to classify clouds into three main categories: cirrus, cumulus, and stratus. Inside those three categories are nearly a hundred subcategories.

You cannot fold a piece of regular notebook or printer paper in half 10 times. The most you can fold it is seven times because once you get to that point, the paper pile is too thick and small for folding.

Roy G. Biv. That's the mnemonic device we're taught to remember the colors of the rainbow: red, orange, yellow, green, blue, indigo, violet. But get this: those are only the colors you can *see*. There are more than seven colors in a rainbow. Your eyes just aren't equipped to see the ones you weren't taught about in grade school.

Everybody knows that glass is actually a liquid that flows very, very, very slowly, right? And everybody would be wrong. Absolutely, glass is a solid, but unlike most solids, its atoms are linked randomly, not in a pattern or a predictable way. Like many other solids, it will act like a liquid when it reaches a melting point.

So, you already know how the universe formed because the Big Bang theory is the official explanation, huh? Well, look at those words again: the Big Bang *theory* is just that: a theory. We don't know the exact, 100 percent correct way the universe was created, and we may never know for sure.

Human Body: Are Zombies Real?

All right, so let's say that a new and special kind of virus hits your area and turns everyone into the living dead. Wait—that can't happen, can it?

To answer that questions, let's start with insects.

Zombie insects, to be exact, which are the unfortunate victims of other, parasitical insects. In the case of the lowly cockroach, for example, the jewel wasp captures the roach, paralyzes it with a first sting, then inserts a type of neural transmitter into its head with a second sting and lets the cockroach go. Eventually, the roach is able to move about, but not of its own free will; it's totally controlled by the jewel wasp, which now has a handy-dandy (and nutritious) place to lay its eggs. The Costa Rican wasp does basically the same thing to the orb weaver spider, only the wasp forces the spider to weave a web for the larvae that are eating the spider alive.

Other insects—such as the carpenter ant and some kinds of cicadas—are felled by fungus, and roly-poly bugs (pill bugs, wood louses, whatever you call them) can be zombie-ized by parasitic worms.

Mammals can become zombies too.

Scientists have recently discovered that a toxoplasma parasite can infect rats, eliminating their natural fears and making them sexually attracted to cats. Naturally, the cats are okay with that in a big way. They'll kill the rat and most likely eat it (or some of it), which is what the parasite wants all along: the parasite needs the cat's digestive system to reproduce. Cat eats infected rat, cat excretes parasite in its feces, parasite finds another rat, and the circle is

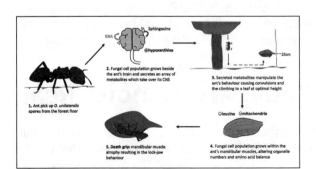

When the zombie-ant fungus Ophiocordyceps unilateralis *infects an ant (often a carpenter ant), it takes over its body quite literally.*

complete. And here's a fun fact: as much as 30 per-cent of the world's population has been infected with toxoplasma.

That doesn't automatically mean anything for you, though. But before we go any further, let's look at the origins of zombies.

Ancient Greeks feared some of their dead and anchored those corpses in their graves with rocks and stakes. Hints of the undead have been found in literature from the very late seventeenth century, but you have to imagine the idea has been around longer than that, right? Film fans have been able to find zombies in movies for almost as long as there have been movies.

When Haitians came to the United States as slaves in the early 1800s, they brought with them dark tales of people who'd died and been brought back to life, often on purpose, by slave masters who wanted a completely obedient worker that wouldn't require food or rest. Zombies originally were tied to voodoo, but after awhile the story morphed into what we know today, that a zombie is created by another zombie who has a virus that makes zombie-hood possible.

And here's where we get a little bit of an an-swer to the original question.

For people who believe in voodoo, zombies are real people who are brought back from the dead by a bokor, or a voodoo practitioner. Sup-posedly, through the use of herbs, shells, and other natural products including tetrodo-toxin, the latter of which is a not-entirely-safe paralytic that leaves its victims unsteady on their feet and in a state of con-fusion, the bokor "creates" a zombie. Indeed, high doses of

Think the U.S. Centers for Disease Control is stuffy? Go to the CDC website and you can download a fictional graphic novel about preparing for a zom-bie apocalypse.

tetrodotoxin can make someone appear as though they've died, which, in some cultures, could result in immediate burial without embalming. It's happened like this: dig up the tetrodotoxin victim, wait until the paralytic wears off some, and voilà! You've got a zombie.

Furthermore, there are some diseases that mimic certain zombie-like traits, including the stiff walk, the all-consuming desire to eat things that aren't really food, the dead look in the eyes, and the inability to do little more than growl. Certain brain diseases, neural damage, and dementia can cause these actions too.

There's good news in all this, though, and it starts with this: the zombies you know from television and movies are probably not something you should fear.

First of all, scientists can offer you hints for surviving the zombie apocalypse (use basic survival skills and a survival kit; pick a secure and windowless hideaway; bring your weapons; stay in a group), but those hints are pretty tongue-in-cheek.

Second—and this is a big one—the standard depth of a burial is about 4 to 6 feet below the sod,

FAST FACT

Early pop-culture zombies didn't go for brains like modern ones do, but it's not a bad idea: they could live on a diet of brains for quite some time, given the water content and vitamins stored in our noggins. The problem would be getting the victim's skull open  in one chomp: zombie jaws aren't very wide, and human skulls are pretty sturdy bones.

> **FAST FACT**
>
> Remember that first stumbling zombie in George Romero's (1940–2017) *The Night of the Living Dead?* His real name was Samuel William Hinzman (1936–2012), and he went on to other roles in other zombie films and later directing two more.

and most cemeteries require a concrete liner these days. That means a lot of tonnage over any kind of coffin that could hold a potential zombie, and it's highly unlikely that the zombie could dig its way out before it would starve to another death.

So, you're pretty safe.

Keep that baseball bat next to the bed, though, just in case.

Human Body: Body Parts You Rarely Use (and Maybe Never Will)

Your body is a magnificent machine. It's also a storage area for things that your ancestors had to have to survive but that you really don't need anymore.

- *Take, for instance,* your appendix. Scientists think that the bodies of foraging ancient humans used the appendix in digestion. Today, because we have more varied and balanced diets, the appendix no longer serves in that capacity. In utero, your appendix helped to preserve your life. In adults, scientists believe that the appendix may still work with the lymphatic or immune systems, or that it's a spot for "good" bacteria. No matter: if you have your appendix removed, you'll be fine without it; in fact, it's a possibility that future humans will evolve without one.

- *Lay your forearm* on a table, palm up. Touch your pinkie finger with your thumb and watch your wrist. Did you see a tendon pop up? If you did, that's your palmaris longus, which helped your ancestors grip tree limbs and climb trees. And if you didn't see the tendon? Don't worry about it; 1 out of every 10 humans evolved without it and live just fine.

Up high under your ribs on your left side, your spleen hides behind a thin tissue that's awfully easy to rip, which could result in the removal of that organ. No problem; lots of people live without one. The spleen regulates the level of red and white blood cells and platelets and removes any cells that are damaged. If you have to have yours removed, your liver will take over the spleen's chores without any complaint.

Your ancestors had big mouths. Literally, their jawbones were wider and deeper, which means they had more room for teeth, such as what we now call wisdom teeth. Your jaw is much smaller than that of your ancient ancestors, and you can eat just fine without those extra chompers. Wisdom teeth, by the way, are another body part that future humans are expected to evolve without.

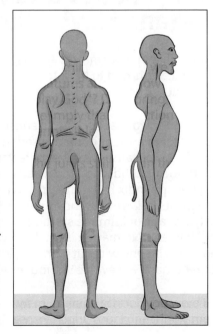

In extremely rare cases, a human can have what's called a vestigial tail, when the bones of the coccyx extend longer than normal and protrude outside the body.

When you get cold, you might get goosebumps or goose flesh. That's the result of the arrector pili muscles, which used to help your ancestors keep warm by making the hairs on their bodies stand up, thus giving them more insulation. You have central heating now, and fleecy shirts. You don't need arrector pili anymore. And by the way, when you get scared and you get goosebumps, it's because the arrector pili are trying to make you seem bigger and fiercer, to repel threats.

Also in utero, male and female fetuses are alike for a certain period of gestation, which means that male babies develop nipples long before the separation of the sexes happens at about six or seven weeks' gestation. Like females, male humans have nipples, and they have breasts (which, by the way, can become cancerous), and it's medically possible for a man to lactate, but doing so is a reason to call your doctor.

Can you wiggle your ears? Those that can have harnessed their auricular muscles, which once helped our ancestors to listen for danger or prey. Just as your dog or cat moves its ears to reflect its moods, humans used to be able to move their ears, but that ability is mostly long gone; even modern-day ear-wigglers can't move their ears as much as their ancestors could.

On the inner corner of your eyelid, there's a small vestige of membrane that used to help protect our ancestors' eyes. It's called the plica semilunaris, or the third eyelid. Some animals still have (and use!) theirs; humans lost the ability to do so eons ago.

Overall, only 2 percent of the world's population has green eyes. Around 80 percent of the world's population has

FAST FACT

Not long after you were conceived, you had a tail. Fortunately, by the time you were a few weeks old in utero, it was absorbed into your developing body, but it didn't disappear. That's why you have a tailbone: it's what's left of your tail from back when your development was most animal-like.

brown eyes. Scientists think that all the people with blue eyes had one single ancestor in common, someone who had a genetic mutation that put less color in his or her eyes.

 Organs you have for a good reason, but can live fine without: reproductive organs, that extra kidney, that extra lung, breasts, eyes and ears, your stomach, and much of your colon. You can also live without both legs, both arms, and a nose, but the loss of those things could be a challenge.

Plants and Animals: Not Going to Leaf Any Time Soon

Who remembers why leaves change color? You probably learned that in grade school and promptly forgot—so just as a reminder, here's what happens from spring until late fall to the trees in your yard.

Surprisingly, to know how trees make leaves, we have to start in the winter! That's when trees are dormant—alive, but just resting because it's cold and resources (such as fertilizer and water) are hard to come by. This doesn't mean they're not doing anything; the buds that formed at the end of the previous summer might have been protected by scales or a kind of edible (by bugs) coating, or they might be naked. Either way, they're waiting.

To break bud, most trees need a cold winter, the length and iciness of which are determined by the species of tree and the area in which it grows. The cold seems to alter the trees' growing process by slowing it. When the right amount of daylight and warmth come back in the spring, the tree is revved and ready to bud with new leaves and new growth. You can bet that global warming can affect this cycle and the ability to grow a tree outside its preferred planting zone.

Tree buds found at the end of a branch are called terminal buds, and they generally have to do with growth of the branch. Axillary buds are buds along the stem at the point of the axil of a leaf, or where leaf meets stem. Axillary buds may become flowers, leaves, or more branches at a later time. If they bud as leaves, these leaves will produce chlorophyll to help nourish the tree via photosynthesis through sunlight. The leaves are green because of the chlorophyll.

Minerals and water help the chlorophyll make sugars that nourish the tree. Furthermore, the leaves help to reduce evaporation from the soil by shading the soil with a canopy that rain can bypass. This also helps cool you and any buildings that the leaves might shade.

As the weather cools and the days get shorter, those changes trigger a breakdown and redistribu-

FAST FACT

Conifers such as pines and cypress trees are green year-round, but they do lose needles on a regular basis, generally in early autumn. It doesn't happen every year, but it happens on a cyclical basis of two to four years or so, over the entire tree.

tion of chlorophyll in the trees' leaves; some of the chlorophyll goes inward, toward the trunk for use later. As chlorophyll is lessened, pigments that were always present but that were overshadowed by the chlorophyll begin to show up. Xanthopyll makes leaves yellow, carotenoids make them look orange, anthocyanins make leaves red. Other leaves turn colors because the available sugar is altered by the weather and less sunlight.

And as the tree's hormones change (led by ethylene) with the season, that triggers changes at the base of the branch, where the leaf attaches. The leaf is cut off from nourishment and it falls, leaving a tiny scar on the branch. If you look at a relatively fresh branch scar, you'll see where the leaf was nourished by the tree.

Some trees keep their leaves for the winter and shed them in the spring. It's believed that doing so is an ancient method of protection, to keep foraging animals such as deer or caribou from eating tender spring growth.

Scientists and gardeners know that there's an easy way to avoid having to rake leaves: *just don't*.

FAST FACT

The General Sherman Tree is located in Sequoia National Park and is the world's largest living single-stem tree. The General Sherman Tree, named after William Tecumseh Sherman (1820–1891), is nearly 275 feet high, more than 102 feet in circumference, and believed to be up to 2,700 years old.

Leaves left on the ground in the fall make great cover for the larvae of beneficial bugs that are waiting for spring just below the surface of the soil. Last year's green leaves make great mulch when there's snow on the ground, and in the spring, they'll suppress the growth of weeds. When it's warm enough, that organic matter breaks down into nitrogen-rich compost, or you can just mow the dead leaves over and they become free fertilizer.

Technology: The Lady Is an Inventor

So far in this book, you've read about a lot of men and the things they've discovered or invented. So here, let's balance the scales a bit.

Let's start with your life outside the house.

It was a snowy day in the winter of 1900 when Mary Anderson (1866–1953) noticed that streetcar drivers had to open their windows to clear the snow so they could see. Sometimes, if the weather was bad enough, the streetcar drivers had to stop the streetcar, disembark, clear the windshields, and resume the trip. Mary went home and invented a spring-arm-loaded wiper with rubber blades that was controlled with a lever inside the streetcar's cab. Twenty-two years later, because driving had become so widely popular by then, Mary's invention was standard equipment on all Cadillac cars.

Officially, Maria E. Beasley (1836–1913) was a "dressmaker," but when the Centennial Exposition opened in Philadelphia in 1876, she became obsessed with the Machinery Hall and visited it as much as she could. This spurred her on to her first invention and patent for a barrel-making machine (1878). She didn't stop there: arguably, Beasley's most important invention was a life raft, which consisted of collapsible parts that made it easy to store and tote. Overall, she held 15 patents for her devices, and none of them had a thing to do with dresses.

A chemist who worked for 40 years at Du-Pont, Steph-anie Kwolek won many honors and was inducted into the National Inventors Hall of Fame.

And then there are the inventions that make our workdays better.

Stephanie Kwolek (1923–2014) had always dreamed of becoming a doctor, so after her graduation from Margaret Morrison Carnegie College at Carnegie Mellon University, she took a job at DuPont with the hopes that it might pay for medical school. Instead, Kwolek went to work on polymer research, hoping to find new fibers that could stand up to extreme use. Her work led to the development of Kevlar, which is used in bulletproof tires and vests.

Dr. Shirley Jackson (1946–), whose research led to such advances as the touch-tone phone and portable fax machine, was the first black woman to graduate from the Massachusetts Institute of Technology, the second black woman in the United States to graduate with a degree in nuclear physics, and the 18th president of Rensselaer Polytechnic Institute.

On the surface of things, it looks like Grace Hopper (1906–1992) had her future figured out when she was young: she attended Vassar and graduated in 1928 with degrees in mathematics and physics; got her master's degree in mathematics from Yale; and taught mathematics at Yale while she worked on her doctorate. She was an associate professor at Vassar when World War II broke out, and she took a leave of absence to join the war effort. Her first assignment was with the Bureau of Ships Computation Project at Harvard where Hopper worked as one of the first few coders on one of the first few modern(ish) computers.

Divorced and with a young son to raise, Bette Nesmith Graham (1924–1980) took a series of random jobs to keep the lights on. In 1951, she landed work as an executive secretary. Working with newly improved typewriters, she was annoyed that with the upgrades, erasers no longer effectively removed mistakes. She tinkered with some paint at home, took the results to work … and in 1958 formed her own company to sell "Mistake Out," a product that was eventually renamed Liquid Paper.

And then there are the inventions that make our homes better places to be.

It's hard to fathom that manufacturers weren't interested in the new invention that Marion Donovan (1917–1998) offered. Like most new mothers she was sick of leaky diapers, and one day she cut her own shower curtain apart and added snaps and folds and an insert that was meant to be used and discarded. Undaunted, and knowing that she had

Accomplished computer programmer and U.S. Navy rear admiral Grace Hopper developed the language that was a precursor to COBOL.

a good thing on her hands, she started her own company in 1949 and began selling her "Boaters" at Saks Fifth Avenue in New York City. They sold out; two years later, she sold the company for a cool million bucks. But disposable diapers weren't the only thing Donovan invented: she went on to collect a total of 20 patents, including one for a new way of using dental floss.

Before Patricia E. Bath (1942–2019) invented her laser-phaco, people who had cataracts resorted to surgeries that were cringeworthy and sometimes not safe. In the early years of her career she worked at the Harlem Hospital's Eye Clinic, bringing surgery to that facility. Four years later, she left to work at UCLA and Charles R. Drew University; next, she moved to Europe and worked in the field of lasers. After she returned to the United States in the late 1970s, she helped found the American Institute for the Prevention of Blindness. In 1981, she began working on her laserphaco as a safer device for removing cataracts; five years later, she received a patent for it.

Imagine a torture device that stands up with stiff whalebone or metal, wraps around the ribs, keeps your spine straight, and pushes your breasts upward in an uncomfortable manner. That was the basic idea for women's undergarments before Mary Phelps Jacob (1891–1970) grew weary of that discomfort and fashioned an easier-to-wear garment with two handkerchiefs and a length of ribbon. You can almost hear the sighs of relief from her family and friends, who wanted one or two for themselves. Jacob patented the undergarment in 1914, calling it the Backless Brassiere and selling it under the name Caresse Crosby; even so, her invention didn't really take off until rationing of metal (the more common material for corset supports) during World War I forced women's hands—and breasts.

Here's another woman who was tired of the status quo: Josephine Cochrane (1839–1913) grew sick of her best dishes and glassware being broken by careless house staff. In 1870, her frustration finally hit its limit, and she began looking for a way to safely, mechanically, and carefully clean her fine china. After the death of her husband, which left Cochran all but destitute, the need to put her ideas to market became more urgent; she began refining her idea,

FAST FACT

As a pioneer in cancer research, Dr. Wilhelmina Dunning (1904–1995) was a researcher at the Crocker Special Research Fund in the 1930s and early 1940s, tasked with studying sarcomas in animals. In order to properly study prostate cancer and tumors, Dunning developed the "Dunning rat" to ensure a controlled base for research.

and by the end of 1886 she had patented a device that used water pressure, rather than abrasive scrubbing, as cleaner. She presented her invention at the Chicago World's Fair in 1893, and voilà!—the dishwasher.

Plants and Animals: Pee-Yoo! That SMELL!

Twelve different species. That's how many different kinds of skunks there are—which is a small bit of information that won't matter to you much if you've ever been on the business end of one of the malodorous creatures. These are the striped skunk, the Eastern spotted skunk, the hooded skunk, the western hog-nosed skunk, the pygmy spotted skunk, Molina's hog-nosed skunk, the Sunda stink badger (that's gotta be the best name), the western spotted skunk, the striped hog-nosed skunk, the Humboldt's hog-nosed skunk, the southern spotted skunk, and the Palawan stink badger. If you wanted to make it easy on yourself, you could categorize them in this way: striped skunk, spotted skunk, hog-nosed skunk, hooded skunk, and stink badgers, which aren't badgers but stink nonetheless.

For now, let's focus on the most common skunk in North America, the familiar, regular old striped skunk, which is found in every one of the lower 48 states and a good chunk of Canada, too.

Let's start off by saying that if you've ever seen a striped skunk, you'd know it. Striped skunks, like all skunks, have coloring that warns predators and mischief-makers alike to *run the other way*. In this case, the striped skunk has a white blaze on his snout and a white V that runs from the top of his head down the back of his neck, across his shoulder blades and across his sides, and plumes down his tail, all of which contrasts perfectly against his shiny ebony coat. Even the most colorblind animal will notice the markings. Those white stripes are unique to the individual, like your fingerprints are to you.

Given his reputation, the skunk isn't very big: he's about the size of your pet cat, at up to 15 pounds (nearly 7 kilograms) and up to 32 inches (nearly a full meter) from tip of the nose to the tip of that long, long tail.

Depending on what's available, a skunk will chow down on grasshoppers, crickets, beetles, or other insects and do some plant browsing, including

The common skunk is found all over North America. It is part of the weasel family and is related to polecats and stink badgers.

FAST FACT

About one in ten people enjoy the smell of skunk. It is the musky component that they enjoy, which, interestingly, in various forms is used in perfumes and colognes. Different types of thiols—organosulfur compounds—from skunk spray are used by the perfume industry to help fragrances last longer.

berries. In the winter, when the insects are not quite as available, the skunk will eat field mice, small birds and eggs, reptiles, and fish. They're pretty fair hunters, and they're not horribly fussy: a striped skunk will eat roadkill and whatever carrion they find, including that left by other predators. They'll also raid a garbage can, if it's easy and smells good.

That means that while you'll find skunks wherever a skunk wants to be—mostly in open and agricultural areas—they can absolutely become urban pests.

Human Body: (Don't) Be a Crybaby

Come here—we'll give you something to cry about! Mere moments after you were born you were able to cry, and you've been doing it your whole life. So let's start with those early cries. The cries you had as a tiny infant differ from the ones you had later when you were older, in that newborns don't produce tears until they're at least two weeks old. Even so, your parents probably didn't notice that tears accompanied your wails because your lacrimal glands didn't produce much of the wet stuff until you were a little older, up to three months old.

For an infant, there are three different kinds of crying:

The basic cry seems rather sing-songy, with crying alternating with silence; researchers say that hunger is the usual reason for a basic cry. Scientists believe that the cadence of the basic cry may lead to speech patterns later. The anger cry is louder and more forceful, and the pain cry is the loudest of all. These may or may not be accompanied by tears.

> Men have larger tear ducts than do women, which may allow them to hold their tears longer than women can hold theirs. Men also have more testosterone, which may inhibit emotional crying.

And about that wet stuff.…

Tears are made by the lachrymal glands, located just above each eye, and are made of three basic materials: mucus from the eye's outer membrane; water, which also washes the eye; and oils from the meibomian gland, which keeps the eyeball from drying out. There are three kinds of tears:

Basal tears are the tears you get when something gets in your eye. They're the tears that move in to protect your eye from irritation. You probably aren't conscious of it, but you produce basal tears almost constantly.

Emotional tears are the ones you shed when you're happy, sad, angry, tired, or otherwise experiencing some feeling or another. Scientists have found stress chemicals and lots of protein in emotional tears, which suggests that such tears are a way to release bad feelings. These kinds of tears release oxytocin and other endorphins too. Often, crying emotional tears—having a good cry—really can be a good thing, and scientists have known that for centuries. Plus, you'll feel better, because we now know that holding back tears is psychologically detrimental.

Reflex tears consist mostly of water and act a lot like basal tears: they appear quickly to wash away irritants.

The reason for the second type of tears has been debated for centuries.

Tears aren't just for conveying emotion, of course. They also help clear irritants away from your eyes. Also, they keep your eyes moist and help them focus light for clearer vision.

FAST FACT

Researchers say that women cry more than five times a month. Men cry less than half as often. As for duration per episode, women cry longer than do men. Also, there are people who don't cry at all, ever, and science is busy studying them. Preliminarily, those people are quick to anger and disgust, and they tend to withdraw more often than their teary-eyed brethren.

Centuries ago, people thought that tears came from the heart and were a product of a weakened ticker, but by Hippocrates' time in the fifth century B.C.E., physicians knew that tears were a good thing, that they were cathartic. In 1662 Dutch scientist Niels Stensen (1638–1686) finally figured out where tears came from, and he had a simple reason for them: they kept the eyes hydrated. He was right, but not all the way.

More modern scientists have other ideas: they've hypothesized that tears may be a way of appeasing a potential enemy, a mode of communication, or a way to express helplessness in the face of anger, pain, or happiness. It's possible that tears may be evolution's way of bonding us too. Tears, because they're so *there*, are viewable by others who are in pain. Tears, therefore, can be a social thing. And then there are the curmudgeons among us, who point out that tears can be manipulative.

The bottom line: don't feel bad if you get that lump in the throat and wet cheeks sometimes, but also know that you can cry too much. Depression, anxiety, and other mental disorders can cause crying jags that cry out for a doctor's help. Be aware.

Plants and Animals: Moss under Glass

If you've ever done a 180 on your ideas and turned away from one thing in favor of another, you're going to like this story....

When Nathaniel Bagshaw Ward (1791–1868) was a mere lad in his teens, he was sent to Jamaica for reasons that have escaped history. It's believed that his interest in tropical plants grew there, but, alas, at that time the only way to transport plants anywhere from continent to continent was by ship, and it was the rare plant that made it alive on such a long and sea-salty journey.

Not much is known about Ward's younger days, but we can assume that he continued to love plants while he reached for higher learning. For sure, by the time he was a physician working in a poverty-stricken East End area of London, he had developed a strong hobby in botany, and he was buggy over entomology, too.

Dr. Nathaniel Bagshaw Ward, inventor of the Wardian case.

Sadly, the neighborhood where he practiced was badly plagued by pollution, fog, smoke, and coal ash, which not only affected Ward's patients but also kept him from cultivating his plants. He specifically became discouraged because he couldn't grow ferns.

He could grow insects, however, and this interest began to consume his free time. One day in 1829, he found the pupa of a sphinx moth and put it in a jar with a small amount of slightly damp soil in the bottom, sealed the jar, and left it to sit.

A few days later, Ward noticed that while the moth hadn't hatched yet, something else was growing in the bottom of the jar: a small fern had emerged from a spore he'd missed in the soil. The jar had condensation on its inside, and he realized what had happened. The humidity level inside the jar was different than what was in the room, and because of that the plant was basically being self-watered.

Like any good scientist with an idea, he began to tinker. What he came up with, he called the "Wardian case," and it caused a minor sensation. Here was a way to cultivate delicate, humidity-loving, warm-

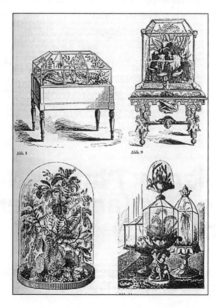

Some examples of early Wardian cases.

FAST FACT

Arguably, the way we eat was influenced by Ward's invention. Without a way to transport actual seedlings (because sometimes seeds don't flourish as well as do cuttings), many fruits and vegetables that we eat today might not be as available as they are.

weather plants in miniature, and they could be moved from place to place easily. What's more, the container didn't have to be small; if a botanist wanted to go large-scale, the only limit was the size of the glass.

It took a few years for the news of Ward's experiments to be published and studied, and the era's best scientists praised his discovery. The Wardian case was good news, not only for scientists but for plant lovers: it wouldn't take long before houseplant mania spread through Great Britain, and everybody had to have greenery in their drawing rooms.

Given his fascination with plants and terrariums, when he died Ward was buried in an unmarked grave in London's historic rural West Norwood Cemetery.

Ward was honored in 1837 when Irish botanist William Henry Harvey (1811–1866) and English botanist Sir William Jackson Hooker (1785–1865) named a South African species of moss after him.

Space Science: Pluto Is Not Just a Cartoon Dog (and Other Planetary Facts)

A mnemonic is a device that helps you remember a list of things in order. For instance, look at the first letter of each word in this sentence: My Very Efficient Mother Just Sent Us Nine Pizzas—which may help you remember the planets in order: Mercury, Venus, Earth, Mars, Jupiter, Saturn, Uranus, Neptune, and Pluto.

And now let's look at those planets and that big bright thing they rotate around.

Mercury

It takes sunlight a little over 3 minutes to reach Mercury; it takes about 5 1/2 hours for sunlight to reach Pluto.

Mercury's daytime surface temperature can rise to 800°F (430°C) because of its proximity to the Sun. But because the planet has no way to keep that heat, the nighttime surface temp can drop down to −290°F (−180°C). Even so, Mercury is not the hottest planet.

There might be a bit of water on the planet, but only in craters where sunlight never reaches.

Lava has been discovered on Mercury, which means that the planet must have volcanoes.

Venus

The surface temperature on Venus can reach around 900°F (480°C). That's hotter than the temperature required to melt zinc.

A preponderance of carbon dioxide in Venus's atmosphere causes the surface pressure to be up to 100 times greater than it is on Earth.

Venus and Earth are roughly the same size.

Most of our universe's planets rotate west to east; Venus and Uranus rotate east to west.

A single day on Venus is equal to 243 days on Earth.

Earth

To put things into perspective, the Sun is 109 times the size of Earth.

Earth is the only planet not named after a Greek or Roman deity. The word "Earth" is Germanic in origin and means "the ground."

The temperature on Earth has a range of more than 270°F. At its coldest, the planet can reach less than −137°F (−94°C) in Antarctica; at its hottest, it can reach temps of nearly 136°F (58°C) in the Middle East. Interestingly enough, humans can survive and thrive in both extremes.

Our planet does not rotate every 24 hours. It rotates every 23 hours and 56 minutes. You've been rounding it up all your life and didn't even realize it.

Mars

If we suddenly had to leave Earth, Mars is the next best thing—although life there would not be easy. The atmosphere, for instance, is mostly carbon dioxide, and the temperature there is definitely colder at about −81°F (−63°C) on average. Still, experiments on soil from Mars proves that vegetable growth there would be possible.

Much of that soil is red or reddish because rocks on Mars are full of iron, which oxidizes into rust.

The solar system officially has eight planets since Pluto was downgraded to the status of dwarf planet. (This illustration is obviously not to scale!)

There is very little atmospheric pressure on Mars, considerably less than that on Earth. Because of this, if you decided to visit Mars without a spacesuit, your body couldn't handle it and you'd die instantly.

Extremely salty water and frost have been discovered on Mars, and the planet may have once had glacier activity.

Jupiter

Jupiter makes a full rotation in 10 Earth hours' time. For comparison, Earth is moving at about 1,000 mph (1600 kph); Jupiter is moving at 22,000 mph (35,400 kph). This speed makes Jupiter slightly flatter on its poles.

If you love birthday parties, don't move to Jupiter. One year on that planet is roughly equivalent to a dozen Earth years.

The very same elements that make Jupiter also made the Sun.

Jupiter has up to 95 moons.

Saturn

Like Jupiter, Saturn consists mostly of gases—although it's possible that Saturn has a small, solid core inside. Still, if you could find a swimming pool big enough to fit, Saturn would float on water like a blow-up pool toy.

Saturn would be a great place to wind surf, if you could stand the cold weather. Surface temps drop to −220°F (−140°C) on average, with wind speeds (east to west, like Saturn's rotation) of up to 500 miles an hour (800 kph).

Saturn's largest satellite, Titan, is the only satellite in our universe with an atmosphere, which makes it more Earth-like than even Mars. Scientists believe that Titan experiences rain and seasons, and it has oceans and rivers. The biggest difference, perhaps, is the surface pressure, which is about 50 percent higher than that of our planet.

Twenty-nine Earth years are equal to one single year on Saturn.

Uranus

You say you want to visit Uranus? Don't wait around: it could take more than nine years for an Earthling spaceship to reach Uranus.

For the record, Uranus wasn't named as a joke. The planet's discoverer, Frederick William Herschel (1738–1822), wanted to name it after King George III of England. Instead, German astronomer Johann Bode (1747–1826) gave the planet its name, after the Greek god of the sky, Ouranos. Either way, the planet's name wasn't widely used until 1850.

The Hubble telescope confirmed that clouds rotate around Uranus at speeds of up to 300 mph (480 kph).

As compared to Earth, Uranus has a nearly 98 degree tilt, causing it to basically lie on its side. Its poles lie east and west and get the most sun; its equator runs top to bottom. This causes some crazy seasonal variations. On the occasions that the Sun shines directly at a pole, the rest of the planet, the part in the dark, experiences a 21-(Earth)-year-long winter.

After constructing his own telescope, William Herschel (1738–1822) discovered Uranus in the spring of 1781. It was the first planet discovered in centuries, and it made Herschel famous. Britain's King George III even appointed him Court Astronomer for his deed.

Neptune

Neptune, Uranus, and Pluto are the only planets that you can't see at night without a telescope.

If you lived on Neptune and you said it had been a long year, you'd really mean it. One single year on that planet is equivalent to 165 years on Earth.

Much of Neptune's atmosphere consists of hydrogen, helium, and methane. The latter absorbs red light, making Neptune appear as a pretty blue marble when you look through a telescope.

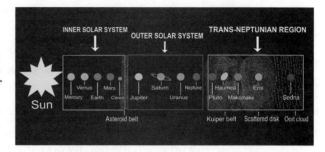

Pluto is a Kuiper Belt object, one of several plane- toids in the Trans-Neptu- nian region.

Neptune's largest satellite, Triton, boasts the coldest tem- perature in the solar system, at –391°F (–235°C). Bundle up!

Pluto

This is a tricky one: Since 2006, Pluto is no longer consid- ered a planet.

The name of this planet was suggested by an 11-year-old British girl in 1930. Venetia Burney (1918–2009) named it after a Roman god.

Pluto is the largest mass in the Kuiper Belt and is nearly 4 billion miles from the Sun. That gives Pluto some small bragging rights out there even though sometimes Pluto is closer to the Sun than Neptune is.

Pluto is small—smaller than many moons in the solar sys- tem—and has five moons all its own. Its temperature could be as cold as –400°F (–240°C).

One day on Pluto is equal to nearly a week on Earth.

Like Jupiter, Pluto has a 122-degree tilt, making it seem to rotate on its side.

Earth Science: A Nice Place to Visit, but Would You Want to Live Here?

Don't let the title of this section fool you: in the widest sense of it, you don't have much of a choice right now. There's nowhere else to go, so you might as well settle in and read these weird things about the Earth.

Let's start with this: we need our planet to remain solid. If Earth suddenly became a hollow sphere like a rubber ball, chances are that the top part would collapse into the middle. You wouldn't care much because you'd be working too hard to hang on when the gravity on Earth suddenly changed. That's okay; the atmosphere would be changed too—so much that life wouldn't last long anyhow. Surface water would evaporate while the oceans would swell and the continents (the parts that didn't flood) would literally become scorched Earth. Plants would die, and animals would follow them pretty quickly.

Okay, forget the hollow planet thing. What if you could make Earth like a glass bead, with a hole through the middle? What if you jumped into that hole?

First, you'd need some protective clothing. Bring a jacket for temperatures that will feel cooler as you fall, but also bring shorts and a tank top, since the middle of the planet hovers just above 9,300°F.

Okay, so you took the leap. The farther into the hole you go, the faster you'll fall, until you reach that super-heated center. The temperature has nothing to do with your slowed descent, but take note of it, because you'll see it again soon: inertia will get you to the halfway mark of your trip through the planet, but you'll fall at an inverse rate. Because the weight of you can't overcome that inverse rate, you'll fall backwards and repeat a little bit of your first trip down, until the weight of you can't overcome the speed of the fall and you'd just go back and forth like that, stuck near the Earth's core.

Again, it won't matter much. You could die at any place along the trip, from the cold, the heat, or

If Earth were truly flat, there would be a number of problems, such as gravity tugging you toward the center and making you walk lopsided.

the pressure. The best thing you could do is hope for 30 seconds to wave goodbye.

Alright. So scratch that little idea.

What if the Earth suddenly flattened like a pancake?

The first thing you'd probably notice would be a problem with gravity, specifically that it would pull toward whatever center there is on a flat plane. This would make you walk in a funny sideways manner, due to the pull; it would also make water head for the center, and plants and trees would grow sideways. Oh, and with no gravity, the Sun and our Moon might not stay where they are in relation to the planet, which could be a bit of a problem: you'd still get days and maybe even seasons, depending on how the Earth-pancake ends up, but with nothing to rotate as a *round* planet does, your calendar could be off pretty quickly—to say nothing of GPS, which wouldn't work because there'd be nothing for a global positioning system satellite to orbit around. That would be okay, though; with that pull toward center, you couldn't really go very far anyhow.

Okay, never mind. What if you could put the brakes on and make the Earth stop rotating?

Let's just say it's going to take a while.

The planet weighs somewhere around 1.317 × 10^{25} pounds (nearly 6 × 10^{24} kilograms), give or take, a number that scientists settled on after studying gravity, rotation, and other numbers that helped them figure things out. The planet rotates at about 1,000 miles (1,600 kilometers) per hour, with one complete rotation in just a tetch under 24 hours. Because of laws of momentum, stopping something *that big* moving *that fast* will take a while.

Even so—and depending on how fast you can get the planet stopped—what's on Earth might spin for a while independently. So, secure your things, as well as soil and oceans, just in case.

Assuming that, like some sort of deranged evil menace, you managed to stop the planet, the first thing you'd notice is that the water in the oceans will move north and south, absent any centrifugal force, which is going to change the look of the continents quite a bit. If you wanted to visit Norway, for instance, better do it before you put on the brakes because the likelihood of Norway being underwater when the planet stops is pretty high.

It'll take you a few hours to notice that your day is suddenly not like yesterday. Without a rotation, Earth's new "day" would be six months long, as would its night—meaning that sunup to sunup would last a year. This might be kind of cool for a little while, but ultimately it would make one side of the planet much hotter for half the (new) day and much colder for the other half. It wouldn't take long for this lengthened day to completely throw the climate off for the whole world, affecting the average person's health and well-being and upsetting markets and demanding cooperation from areas that can grow food. This means that you *could* live, if you and the planet's crops and livestock could withstand the extreme temps on an Earth that doesn't rotate and is suddenly full of hard winds.

The good news is that you wouldn't be flung off the side of the sphere; there would still be enough gravity to keep you on the planet.

FAST FACT

We know the correlation between geology and meteorology thanks to William Morris Davis (1850–1934). His first passion for study was engineering, a subject for which he received a master's degree from Harvard. He also had a fascination for geology and geography, as well as meteorology, and he posed the idea that valleys and plains were formed and shaped by such things as weather and water. Today, Davis is known as "The Father of American Geography."

Plants and Animals: Menagerie, Part 4

The Scottish Natural Heritage has in place a plan should the Loch Ness monster ever be captured alive. They'll take a DNA sample or two and release the creature back where it was found.

When a camel drinks a large amount of water (and they do—up to 20 gallons or 110 liters at one time!), the H_2O is stored not in its hump but in its bloodstream. This dilutes the animal's blood to the point that would kill almost any other animal.

Scientists believe that a single mass extinction more than 250 million years ago destroyed more than 90 percent of the creatures on Earth.

As a horse ages, its gums begin to recede, and its teeth look as though they've grown. This is where the old saying "long in the tooth" comes from: the older the horse, the longer the teeth appear to be.

Snakes are strictly carnivores; they do not eat plants at all because they don't possess the gut bacteria to digest them.

Among all the bruins, the polar bear has the largest paws. Those dinner-plate-size feet help the polar bear walk easily on snow and ice.

Peter Rabbit's creator, Beatrix Potter (1866–1943), obsessively studied mycology when she was not book-writing. She came up with a theory about fungi reproducing by way of spores, but because she was a woman and young, her work wasn't perceived as well as that of men of the time. In 1997, the Linnean Society of London, which is a hub of information on natural history and evolution, issued a posthumous apology to Potter.

Reports suggest that more than 45,000 people die each year in India of snakebite. Around 80 percent of the world's deadly snake bites happen in India.

Recent studies show that dolphins recognize their friends by a series of vocal sounds, and by taking small sips of their urine.

Human Body: Brains! ... Brains!

Don't brag: you might believe that *Homo sapiens* have impressively large brains, and we do ... as compared, say, to a mouse, a dog, or a raccoon. But brain size has nothing to do with *intelligence.* Now, granted, intelligence is a relative thing; a mouse is smart in mouse things, a kangaroo is smart in kangaroo things.

When you are on a roller coaster ride, your internal organs—your brain, stomach, intestines, liver, lungs, and heart—are also sloshing around inside you. Likewise, when you feel that stomach flip-flop of free-falling, your organs are affected by gravity and will push one another around inside your body.

It is entirely possible to live a normal, productive, happy life with very little brain matter in your noggin. Science journals are full of stories of adult humans who report to a hospital with minor symptoms, only to learn that their heads are full of undrained fluid and a small percentage of neurons.

 Roughly five percent of the world's population suffers from musical anhedonia, which is a neurological condition that leaves people with a complete inability to receive pleasure from listening to music.

Notable Names: David Hahn

❚❚Do a Good Turn Daily."That's America's Boy Scouts slogan, and it can be as simple or complex as the individual wants it to be on that day. In a way, that's what David Hahn (1976–2016) did: he proved that there was a gaping hole in our country's nuclear security.

When he was a boy growing up near Detroit, Michigan, Hahn split his time between his father's home and that of his mother and her boyfriend. His childhood was like that of many other Midwestern kids: baseball, soccer, exploring, playing with friends, and joining a kids' social group—in Hahn's case, the Boy Scouts.

When Hahn was 10, his step-grandfather gave him a book on chemistry, and the boy was absolutely smitten by the subject. He devoured that book, performing all the experiments inside it; by age 12, he was reading and understanding his father's college-level chemistry books. He purchased laboratory materials and set up a simple lab in his father's house, supplying himself with more guidebooks and textbooks. He couldn't get enough of the subject of chemistry, and though his parents supported his curiosity, they reportedly weren't happy about the explosions that came from his room.

Yes, young David had learned to make substances that went beyond the scope of a normal teenager's interest. Like nitroglycerin. He held a few typical teen jobs so that he could learn to make things like that. He thought that radiation was the key to fixing the energy crisis.

His teachers loved him. Other adults—not always so much. He was prone to telling people what was really in their food and blowing up things that weren't meant to be blown up. His parents began to worry, and they began confiscating some of his experiments when he was at school, terrified that he might leave a crater where their house should be.

Looking for a way to channel his son's interest, Hahn's father encouraged young Hahn to work toward becoming an Eagle Scout, a feat that's notoriously challenging to attain. Hahn was, of course, most interested in earning the Atomic Energy badge, and he achieved his goal in early 1991, just a few months before his 15th birthday. This was just before he almost blinded himself in the family's basement, and his laboratory was banished to his mother's potting shed in the back of the back yard. She and Hahn's father thought it strange that their son spent late nights out in the shed, and they noticed that he was wearing a gas mask out there sometimes, but at least he was away from the house.

By this time, his relatively small focus in one scientific branch sparked a deeper interest in things that most young teens don't think much about: David Hahn decided to see what he could do with radioactive materials.

Hahn set up a handful of pseudonyms and matching stories in order to get what he needed from places that supplied him. He implied that he was a chemistry teacher at one point, and he masqueraded as a scientist. He set up a Geiger counter in his car—one he had ordered from a regular mail-order catalog and put together himself. Large entities—all of them government-vetted, including the Nuclear Regulatory Commission—happily sent him printed information and basically told Hahn where he could get, find, beg, borrow, or otherwise procure the raw materials for what he was building. Some of the materials were inside everyday household items; he just had to find enough of those items and remove the materials.

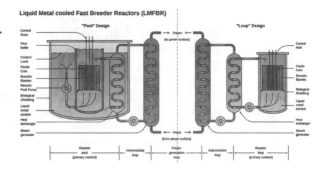

Two types of breeder reactors are shown here. Breeder reactors can use common isotopes such as uranium-238 or thorium-232.

Still just a teenager, he decided to build a breeder reactor.

It should go without saying that real, college-educated, über-experienced adult scientists had tried to do that kind of thing before, with the help of other college-educated, über-experienced adult scientists, and failed. That didn't stop young Hahn.

He knew that if he didn't have enough enriched uranium, there would be no chain reaction like there would be with a large nuclear bomb, but he had enough to mess around with—and he did, until it finally occurred to him that throwing away his clothing at the end of each evening in the lab and wearing a lead apron were probably not enough to be safe. Not just for him, but for his neighborhood too, and so he began to take his lab apart. He put the parts of his reactor in the trunk of his car.

In late summer of 1994, police stopped Hahn on suspicion of tire theft, and when he opened his trunk for them, the officers involved became alarmed. The trunk contained a toolbox with powder and ores and vials inside it, and it made them nervous—especially when Hahn told them that the toolbox was radioactive.

Simply put, a breeder reactor is fueled by the irradiation of uranium-238, which it also uses to generate energy. A breeder reactor uses power but generates more power than it uses.

The Federal Radiological Emergency Response Plan was activated. The FBI got involved, as did the Department of Energy and several other government departments. Nearly *two months later*, after dragging his feet with authorities, Hahn admitted that he had a laboratory out back. His mother had already thrown a lot of radioactive utensils into the garbage, but enough tools and containers were found to indicate a really big problem.

More government departments got involved, and it was determined that there was a possibility of a radioactive leak near the shed and the land around it. Nearly a year after Hahn was stopped by police and his car was confiscated, the shed in his mother's backyard was declared in need of a Superfund cleanup. Close to 40 sealed barrels containing what was left of Hahn's boyhood experiment ended up in the Utah desert. A determination was made that there was no damage to the trees or plants in the neighborhood, but the 40,000 people in the subdivision where Hahn lived might have been affected by the radioactive dust.

Though he was (understandably) distraught over losing his project and all his materials, Hahn managed to land his Eagle Scout status, but not without a fight. The rest of his high school years were tainted by teasing; after graduation he enrolled in college but didn't do well there either. Hahn joined the military and continued his habit of immersing himself in scientific subjects that most interested him. In 2007, after leaving the U.S. military, he was suspected of dabbling with raw radioactive materials again and was arrested for theft of some smoke detectors, which hold americium, a substance in nuclear reactors.

Proof that men didn't have a corner on working with nuclear science: Wilma M. Pickett worked in Los Alamos with the Women's Army Corps at the time that the Manhattan Project was in the works.

As a young man, Hahn was offered the chance at a

Earth's iron core has a diameter of 1,615 miles (2,600 kilometers) and has a temperature of about 9,800°F (5,200°C).

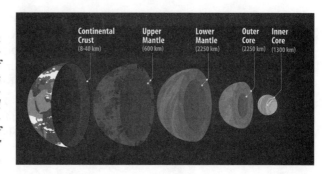

more-than-military-grade, very thorough examination to determine how the radiation he was exposed to affected his body. He declined, saying he was sure that it hadn't stolen more than a few years of his life.

In a way, he was right: Hahn died at age 39 from drug and alcohol use.

Earth Science: Down to the Core

It's been recently discovered and confirmed that the center of the Earth's outer core is an inner core made of 400 miles of solid iron and nickel metals. Researchers believe that this inner core is only about a billion years old and that it has something to do with the planet's magnetic field.

In 1883, the volcano Krakatoa erupted so loudly that the sound of it could be heard 3,000 miles (4,800 kilometers) away. So dangerously loud was the explosion that sailors in the ocean 40 miles (64 kilometers) from the eruption suffered from burst eardrums.

Human Body: You Are Weird

You know the back of your hand like … well, like the back of your hand. But there are probably a lot of things you don't know about your human mind and body.

Your senses can get mixed up and work crosswise. For instance, if you read the color "red" but it's printed in green, your eyes may see the color before the word. In the same way, your eyes can make you hear words to match a video rather than words that are actually said; you can taste things before you actually put them in your mouth; and what you see can challenge your sense of temperature, softness, or hardness. Or try this: next time you have an itch on your arm, look into a large mirror and scratch the same spot on the *other* arm. Chances are, you'll get relief from the itch, despite that you didn't scratch the actual itch.

You may think you can focus on two things at once, but studies show that if you're interrupted in a task to do another task, you may suffer from "attention residue," which will keep you from being fully focused on either thing.

"Highway hypnosis" is another way your brain tricks you. Highway hypnosis happens when your brain experiences a "procedural memory," which is a memory of something you're so used to doing that you barely have to think about it. Your body reacts without fully engaging your brain in the task. Hence, you drive to work, for instance, but you hardly remember the trip.

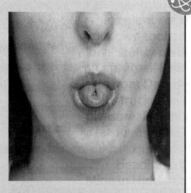

Can you roll your tongue into a tube? Up to 80 percent of us can, with more women on the can-do side than men. Up to 20 percent of us can wiggle our ears. A little more than 3 in 10 people can wiggle their nose.

Good luck telling yourself not to think about something because it's a sure way that you will, due to the "ironic process" theory. For instance: right now, *don't think about a green rabbit*. You're welcome.

You can literally live with just half a brain. A surprising number of people do, and don't even realize it.

Your skin cells developed in a line called Blaschko's lines, which are invisible on most people—which means, yes, your body is covered in stripes you can't see. For those whose Blaschko's lines are overt, the visibility is usually the result of one of many non-life-threatening diseases.

Most people have three creases on their palms. About 1 in 30 have a "palmar crease," formerly called a "simian crease," which is when a crease runs completely from side to side of the palm, clear across from beneath the pinkie finger to between the pointer and thumb. You didn't ask for this crease; it's something that developed in the womb. More men than women have it, and though it's completely normal, it may indicate issues such as Down syndrome.

We like to think we're a hundred percent unique and special—and you are, unless you're an identical twin. In that case, because you were born from a split egg containing the exact same sperm and egg from your parents, you share all your genetic makeup with your identical twin. Furthermore, if identical twins marry identical twins, all of their offspring are genetically comparable to siblings even when they are cousins. Having said all that, you and your twin still

Famous gangster John Dillinger once tried to have his fingerprints removed, but he found out you can't just sand off your prints. They are forever embedded in your fingers.

have physical differences, and you're still unique and special.

⚛ *Famously, gangster John* Dillinger (1903–1934) tried to file his fingerprints off the tips of his fingers to evade the law. Obviously, he wasn't one of a handful of people who are born without fingerprints, due to a genetic mutation known as adermatoglyphia.

⚛ *Most people's hair* whorls at the crown (near the top) of your skull in a clockwise way. You may have two parietal whorls. Three is not unheard of, but it's pretty uncommon. Studies have suggested a correlation between parietal whorl direction, handedness, and sexual orientation, but nothing has been definitively determined yet.

⚛ *About 10 percent* of humans are left-handed. Thirty-five percent of gorillas and chimpanzees prefer to use their left hands. This may have to do with communication and gesture-as-language.

⚛ *While there's a* bit of research to be made yet, scientists believe that it's entirely possible—some say even likely—for you to have a doppelgänger, or a virtual double—someone who's not closely related to you but looks like you enough to startle people. It's rare, but it happens. What science *does* know suggests that doppelgängers often share some DNA and often have similar lifestyles. If you're curious about where your doppelgänger is, there are websites to help you find him or her (because sometimes, it's the opposite sex!).

FAST FACT

When Willie Lincoln (1850–1862) died, the art of embalming was in its infancy; still, his father approved of the method of preserving a body for a little longer before burial. Three years later, when Abraham Lincoln (1809–1865) was assassinated, the doctor who embalmed Willie also did the same for his father.

And finally, scientists know that an earworm, or that bit of song that runs through your head a million times until you're ready to beg for it to go away, is caused by a glitch in the auditory cortex, causing it to want to fill in a gap in the words to the song. It's as if the brain's record player needs a new needle (remember those?). For help, go find the lyrics and read them as you listen to the song. That ought to get rid of your earworm. Chewing gum is also said to work.

Human Body: Urp!

What goes up must come down, says physics. Ugh, and the opposite might include dinner.

So let's say you ate too much. *Waaaaaay* too much, and that was absolutely *not* a good idea. Maybe something didn't "agree with you," or you had a hidden allergy, or you had a virus you didn't know about, acid reflux, morning sickness, an infection somewhere, or gastroenteritis, or motion sickness. Or maybe you just ate too much. Any way you look at it, you start to think that maybe finding a large basin might not be a bad idea.

The first thing to know about vomiting is this: you vomit in order to rid your body of something

The need to vomit is triggered by the chemoreceptor trigger zone in the brain's medulla oblongata, which is the oldest part of the brain, responsible for survival instincts and gut, shall we say, reactions.

harmful or irritating to your stomach. The body believes it's been poisoned, and the fastest way to rid itself of that toxin is to upchuck it. Like most people, you might even hate to vomit, but the fact is that, with a few notable exceptions, vomiting is your body's attempt to save your life. Can't be mad about that.

Back to that meal you overconsumed: when you start to feel urpy, that's a good sign that your body has already sent a message to a part of the brain called the CTZ, which stands for chemoreceptor trigger zone. The CTZ is in the fourth ventricle of your brain, and that's where the decision is made—have you been poisoned or not? If the answer is yes, real or not, then the CTZ sends signals to the rest of the body to prepare for the inevitable. You'll start to get sweaty and your salivary glands will go into overdrive in order to protect your teeth from the stomach acid bath that's coming....

You'd best head for the bathroom.

Immediately before you vomit, all the muscles between your neck and your stomach, including the diaphragm, your chest muscles, and your abdominal muscles, will contract to force the stomach to empty itself upward. A couple of choking dry heaves, the epiglottis shuts to prevent partially digested food and stomach acids from entering your lungs and windpipe, the stomach keeps contracting, and it all comes out where it went in.

Once you're done and your stomach is emptied, unless you've got something long-term going on, you might immediately feel better. That's your body at work: your blood pressure likely falls back to nor-

Science says that motion sickness is caused by a mix-up between the eyes and the brain. It's thought that the brain knows your body is relatively inert but the eyes tell the brain something else. To be safe, the CTZ starts the vomiting process, in case you've been poisoned.

mal, and you may feel an odd rush of endorphins after such a stressful event. That's not to say that you can go straight out and feast again, though: vomiting causes your body to lose electrolytes and fluids, so you need to drink something, even if it's in sips. You should also rid your teeth of what's left of the stomach acids they were just exposed to.

Then take it easy, and be thankful that you're not a rodent, rabbit, or horse. Scientists have only recently discovered that their brains and bodies aren't adapted to allow them to vomit.

Plants and Animals: Up the Waterspout

Poor Little Miss Muffet. There she was just minding her own business, having lunch, when she learned that, like up to 15 percent of the U.S. population, she suffered from arachnophobia.

If she had stopped screaming, she would have learned a little bit by watching that spider who frightened her away.

She would have noticed that, like arachnids such as scorpions, mites, ticks, and others, a spider has a body divided into two sections, called tagmata: there's the head part and the body part, called the opisthosoma. Also, like other arachnids, the spider has eight legs, no antennae, and no wings; it's a predatory creature with chelicerae instead of jaws, and it likely begins its meals by injecting a digestive fluid into whatever prey is caught. Spiders also have specialized fangs, walking legs, eyes, and an exoskeleton that it occasionally sheds.

The differences between spiders and other arachnids are as varied as their similarities. Spiders are capable of making web silk, first of all. Unlike other arachnids, spiders can carry venom, which they transmit through bites. And their mating processes differ from their arachnid cousins: in many

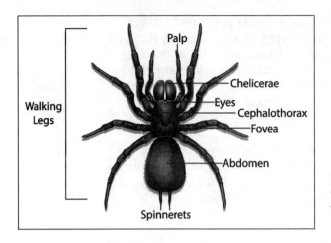

The basic anatomy of a spider.

groups, male spiders transfer sperm packets to the female via special legs called pedipalps during mating. Pedipalps are not only for fun and games, though; they also are used in snagging prey.

If Miss Muffet had dared to get closer, she would have noticed that spider legs have seven segments, ending in teensy claws. Web-spinning spiders have three claws, while hunting spiders generally have just two claws. At least one—the Australian crab spider—has no claws at all. The claws, when they have them, help the spider to grasp and hold.

Generally speaking, spiders have eight eyes, each possessing a single lens—although the number of eyes can vary, and some spiders have no eyes! Some species of spiders have very good night vision; hunting spiders see in color and have pretty good vision; the spiders you notice in your garden or hanging out on a web on the porch probably have poor vision and rely on wind, air pressure, and vibration to "see" their prey. Chances are, then, that they would have only sensed Miss Muffet's escape.

At the end of the opisthosoma, beneath the anus, most spiders have a cluster of between one and four spinnerets, which produce silk for their webs. Thousands and thousands of dry silk threads combine to make the product, which ends up being

0.0001 inches (0.003 millimeters) thick, or about a twentieth the width of a hair on your head.

Though spiders have been making webs for more than 100 million years, evolution has made the kind of web you clean off your shed door: Today's farsighted spider's webs feature scattered sticky dots of liquid that keep small prey trapped until the spider can climb down to immobilize the creature and invite it for dinner—prey that can range from small insects to birds or bats, in the case of larger spiders. The spiders with better eyesight, as mentioned above, generally don't build webs at all, but instead prefer to hunt without.

Spiders will eat their own webs when they need to make repairs. The proteins in the silk are nutritious for the spider.

Everything you've heard about spider digestion is true: when prey becomes entrapped in a web, the spider first immobilizes the prey and then injects a digestive fluid into the victim, which is probably exactly as horrifying as it sounds. This fluid partially digests the prey and makes it possible for the spider to then consume the liquid as meal. If there's a wealth of prey on a web, that's fine too: many spiders will keep potential meals alive, bound in silk, until they get hungry enough. Most spiders won't start the digestion process until they're ready to eat.

> Cobwebs are just spider webs that have been abandoned. The word comes from the archaic *coppe*, which was once a word for "spider."

FAST FACT

Spiders can travel on their own web because they have specialized feet and because they can make both sticky and non-sticky strands. The spider can then carefully travel all over on the non-sticky bits.

FAST FACT

In ancient times, patient artists with steady hands used the webs to both paint on and with. Spider silk, which contains a lot of vitamin K, is often used in folk medicine to stop bleeding. Pharmacology sometimes uses spiders to test drugs, relying on the web-building to see how the substances alter the web. Spider silks have been used in gunsights, telescopes, violin strings, fiber-optic cable, protective gear, and clothing, among other things.

Like everybody else on Earth, though, some spiders gotta be different.

Some use their webs as nets to catch insects or prey as large as fish. Some don't spin a web, preferring instead to snare prey with a single thread. Some build a funnel, eating whatever falls down to the bottom. And some spiders use their web-spinning talents to make a balloon of silk and float to a new home.

Technology: Talk to the Keyboard

Artificial intelligence (AI) is a field of computer science that aims to create systems that can perform tasks that normally require human intelligence, such as visual perception, speech recognition, decision-making, and language translation.

AI systems are typically built using machine learning, a subfield of AI that involves training algorithms on data to enable them to make predictions or decisions without being explicitly programmed. There are different types of machine learning, including supervised learning, unsupervised learning, and reinforcement learning, each with their own strengths and weaknesses.

AI is transforming many industries and has the potential to solve complex problems and improve our lives in countless ways, but it also raises ethical and societal concerns, such as the potential loss of jobs and the potential for AI to perpetuate existing biases. As a result, it's important for college students studying AI to also study the social and ethical implications of the technology.

...

Not a bad couple of paragraphs, eh? Simple to understand, easy, and to-the-point. But get this: the above was taken, word for word, from an AI bot at ChatGPT.com.

Yes, a computer wrote that entry.

To many people, that's scary: the idea that a computer could communicate with humans so easily and succinctly gives many people pause; that they can do it in such a breezily *human-like* manner is almost a little shocking. It's even more shocking that many AI experts admit that they don't totally know exactly how this works.

With the common use of AI, many issues have surfaced of concern to many people—one being that computers are fully capable of "hallucinating," or simply making things up to get by. It's also easy

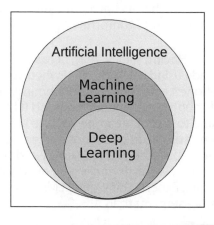

AI is increasingly using deep learning, a type of machine learning that uses algorithms meant to imitate human brain processes.

to replicate just about anything through AI, including images and language, to make it look or sound as if someone has said something they very much DID NOT say. Because it's hard to tell the difference, this can fool a lot of people, especially the gullible. And then there's this: there have been many examples in which two or more linked computers have created their own language and have communicated with other computers without the input *or knowledge* of their programmers. Computers, in other words, are starting to act like us.

The thing to remember above all that is that computers are not sentient beings, they don't yet know what's true and what's not, and they can't foresee any kind of future. They only know what their human keepers tell them—and yet, they are capable of learning through analysis of millions of bits of data that they can quickly sort through. They only *seem* sentient (defined as having the power of perception, or conscious) because they are taught to mimic human speech and speech patterns and because they are quicker than we are in answering questions (which, admittedly, makes them great for fact-checking, but with the caveat that AI can be terribly wrong just as easily as it can be correct). That ability to winnow through data, quickly, is why a computer will often give you two different answers to the same question and why it seems so creepily alive: it has learned to improvise on the fly, if you will. For this reason, what a chatbot may tell you today will likely change tomorrow.

Having said all that, here's the thing: you're already using a version of AI when you use a smartphone's autocorrect programming. That's done through large language models (LLMs), which are able to "learn" the more you use them. That explains why autofill on your phone often guesses the wrong word: because you've used a word paired with another word, it "thinks" you must want to do it again. Experts think that LLMs may someday seem more sentient than AI.

Still, and for all of these exact reasons, we should be thankful that most current chatbots are made to avoid engagement with illegal or harmful questions asked of them.

As for the above brief article, though this is a very basic explanation (the system was asked to explain AI like a college student would), it's not bad in content and grammar, but the tone of the text is not at a level expected from a college student. There's no doubt that AI will completely change the way we read and write, and—according to experts in the field—how we learn.

Notable Names: The Scientist and the Pea—Gregor Mendel

Scientists are already well aware that things like eye color and natural hair color are inherited from the genes we get from our parents. They may have been able to deduce something like that through observation, but it's something we *know*. And we know it because of one man's efforts with simple plants.

Gregor (né Johann) Mendel (1822–1884) was born in what is now the Czech Republic, the middle child and only son of farmers who'd been working

Gregor Mendel was an Augustinian friar and abbot, as well as a mathematician, meteorologist, and biologist famous for his theories on inherited traits.

the family land for several generations. As a boy, young Johann learned to plant and sow and contributed to the family larder through beekeeping, but farm life was not his destiny: a priest recognized Johann's intelligence, and the boy was sent away to school at age 11. Two years later, he was sent on for more education, but he struggled with homesickness and depression, possibly brought on by a lack of support from his family, who didn't have the means to help him. Twice he was so ill that he was sent home to recover, only to return and continue his education.

Nobody could accuse him of a lack of determination: when it became clear that the family could no longer financially support his schooling and that, as his father's only son, he was supposed to take over the family farm, Johann chose another path. He became a monk by joining the Order of St. Augustine. In doing so, he would not have to pay for college.

It was there that he was given the name Gregor.

The monastery was both good and bad for Mendel: The financial struggles he'd endured were over. Also, for the first time in his life, he was surrounded by highly educated people in a diverse situation. On the other hand, his monasterial work of caring for the sick and indigent upset him to the point that his depression returned. Mendel was then assigned to be a teacher, and he was very good at that, but he couldn't manage to pass the tests he needed to take to become certified. In 1850, he was sent to the University of Vienna for more instruction, this time to learn how to teach science. Mendel spent all his time studying math, physics, and botany through the use of a micro-

Peas have been eaten for thousands of years and may have been enjoyed by Neanderthals. Because they're easy plants to cultivate, dry, and store, they were likely one of the first crops planted in North America when the Europeans first arrived.

scope, and he worked with some learned men there. Three years after arriving in Vienna, he returned to the monastery, where he was assigned as a teacher to another classroom.

But something must've stuck in Mendel's head, and in 1854 he received permission from his abbot to organize a large experiment in hybridization, with the end goal of learning how certain traits are carried genetically from generation to generation. Previous experiments done by others had suggested that hybrid plants reverted to original characteristics, so it didn't seem that hybridization would help yield better results in harvests—but then again, farmers had proven time and again that breeding different kinds of cattle or breeds of sheep together made different, sometimes better, strains of cattle and sheep.

For this project, Mendel picked the common edible green pea to study. From 1854 to 1856, he repeatedly bred and cross-bred 34 strains of peas together, looking for seven different traits that the plants would share or not. Mendel then took his findings further by relating them to cell theory of fusion, which states that a new organism springs from the union of two cells, but only when one cell allows the other cell's traits to dominate. Dominant and recessive genes were the ticket to how traits are passed down.

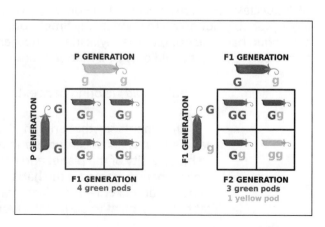

Mendel figured out how genes for certain traits come in dominant and recessive forms.

In 1856, Mendel tried for the teacher's certification again and failed, which led to another bout of depression. Fortunately, he recovered and the head of the monastery apparently didn't worry too much about it: the position Mendel held was his for another 14 years and led him to be elected abbot at the monastery. There, his role kept him from further experimentation; instead, his focus was needed on running the monastery and fighting with authorities on a tax that Mendel felt was unfair.

Though he'd given two talks on his findings in 1865, Mendel made few copies of his lectures and didn't otherwise feel an urge to publish his findings. In the early 1870s, he studied meteorology and harked back to a childhood love of beekeeping, but for years after his death, Mendel's work went largely ignored. It wasn't until slightly into the twentieth century that scientists began slowly using his work to support or enhance their own. Today, Gregor Mendel—failed teacher, intellectual monk—is known as the Father of Genetics.

Plants and Animals: Should We Bring Back Extinct Animals? Part 2

We have already determined that bringing back some extinct species is possible, at least a little bit. We know *Jurassic Park*-ing is probably not going to happen any time soon, but if science can't bring back the dodo or the thylacine or another recently extinct creature *now*, it could only be a matter of time until it can.

But *should* we bring back extinct animals?

Proponents say that bringing back creatures that died out years or centuries ago could increase the biodiversity of the planet and could help reverse climate change, based on the habits they believe an extinct creature kept—if, for instance, a mammal's scat helped a plant to germinate, wouldn't it help that plant's descendant now? Bringing back a creature that died out because of human folly could be righting a wrong if forethought is included in choosing the animal to be resurrected. There's also the possibility that a further mix of DNA with existing animals could prevent future extinctions.

The louder voices—the most vociferous opponents of bringing back extinct creatures—have many good points, starting with money. It ain't cheap to mess with DNA, first of all; getting and using the best technology in the world isn't a matter of pocket change, and the inevitable early failures aren't free. Then there's the expense of securely and humanely housing and feeding the animals that could result from de-extinction. If the creatures' lives depended on a particular kind of food that is also extinct, are we playing on a slippery slope? If we resurrect an animal whose habitat is also extinct, where would we put it?

And though it's only a movie, didn't we learn enough from *Jurassic Park*?

Furthermore, scientists say that though it's tempting to want to see a living, breathing saber-tooth tiger or a dodo in the flesh, the money spent on bringing back those animals is much better spent on conservation for the animals and birds that are around now, that are not extinct or anywhere near it, and that have a good chance of remaining plentiful.

And then there's the issue of morality. Does bringing back an extinct animal destroy the natural

Is it wise of us to try to resurrect extinct species like the Dodo?

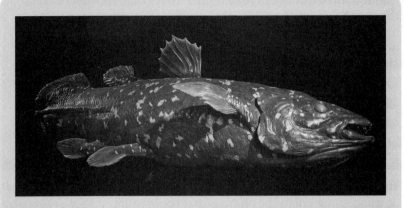

The Coelacanth is a fish that many refer to as a living fossil. There are two species of coelacanth living today (*Latimeria chalumnae* and *Latimeria menadoensis*), having survived in our oceans since the Devonian Period some 400 million years ago. It was believed to have gone extinct 65 million years ago around the same time as the dinosaurs but was rediscovered in 1935 off the coast of South Africa.

balance of things? Would we be messing with sentient creatures and unknown outcomes? Is it up to us to "play God"? Does bringing back an extinct creature imply that extinction is reversible, therefore, permissible?

The jury's still busy in the laboratory. Stay tuned....

Human Body: Come to Your Senses!

 Since you were in grade school, you were told that you have five senses: sight, touch, hearing, smell, and taste. The truth is that there are at least seven that scientists now recognize (the above five, plus vestibular, or balance; and a body-awareness sense, meaning that you know where your body parts are). Despite these two new senses, it's believed

> **The five basic tastes are sweet, salty, sour, bitter, and the recently added umami, representing the savory taste that comes from glutamates and nucleotides.**

that a mere seven may still be a count that's too low: some scientists claim that we may have as many as 33 different senses, depending on how you want to define "senses." Oh, and by the way, those "maps" of your tongue that show where you taste different things? That's been debunked.

Proof that your ears can kill you: sounds above 170 decibels can cause pulmonary embolisms and burst lungs. For what it's worth, your average rock concert hits as much as 120 decibels, which is not good for your hearing, but it won't kill you.

You know those memes that go around once in awhile, like the "is the dress gray or green" one or the "what color are the shoes" one? Your perception of its color is different from the perception held by your sibling, best friend, and neighbor, because of microscopic differences in cones in the eyes and because of the messages the brain gets. A difference in genetic makeup and culture may also factor into what you see.

Plants and Animals: A Funeral for an Ant

According to scientists—who are really good guessers, in this case—there are approximately 20 quadrillion (that's a 20 followed by 15 zeroes) ants in the world, more than 2 million ants for every single person on Earth. It takes about 450 ants to lift one pound, so it would take about 78,000 ants to lift the average human

A million ants would weigh about as much as a half-gallon of paint. An African driver ant queen can lay three to four *million* eggs every 25 days of her adult life.

(theoretically). The 2 million ants could lift 25 people, and 20 quadrillion could easily lift all the people on Earth!

Some species of queen ants can live for decades. Most male ants die a matter of days after hatching. Regular worker ants, like the ones you see on the sidewalk, are female ants in a nonbreeding stage of adulthood, and they could live between a few months to a couple of years, depending on the food they can get and the general health of the nest.

Among the largest ants on Earth are the Matabele ants of sub-Saharan Africa. If one of these is injured in battle, its comrades will pick her up, return her to the nest, and she will be taken care of, much as a soldier is cared for by a nurse. If she dies in the nest or along the sidewalk, she will probably just lie there where she fell. Once that happens, another ant would be spurred, biologically, to pick up her body and carry her to a midden, which is literally a dumping place for dead ants, thus helping to keep disease and bad bacteria from the nest. If moving a dead ant out of the nest isn't possible, the undertakers will move the corpse to a safe, sequestered place in the nest.

Some species of ants even have graveyards in which dead ants are buried.

In an experiment, biologist Edward O. Wilson (1929–2021) proved that if an ant somehow gets the death scent on itself by accident (or experiment), the living ant will be summarily carried by a nestmate, without fanfare, to the graveyard. After it's dumped, it will wash itself thoroughly; if it removes enough of the death scent, it's generally allowed to resume its place within the colony. If not, it gets dumped again, to take another good bath.

FAST FACT

Much of what we know about the social lives of ants, we know thanks to William Morton Wheeler (1865–1937). Wheeler was sequentially a professor of zoology at the University of Texas, curator at the Museum of National History, and professor of applied biology at Harvard, and was a leading authority on social insects. He was one of the first people to discover the complexity of ant society.

This need to carry the dead away, say pest experts, is why you shouldn't kill ants in your home or squash them just outside your door: dead ants emit scents that urge other ants to come around to see why those smells exist.

Earth Science: Mary, Mary, Quite Contrary, Part 2

Sometimes science is a really deep subject. Take, for instance, the Mariana Trench, sometimes known as the Marianas Trench, which is the deepest spot in any of the Earth's oceans and the deepest point on Earth. If you were looking for the trench, you'd find it east and south of the Mariana Islands in the Pacific Ocean.

Created when one of the planet's tectonic plates collided with and slid beneath another, the Mariana Trench is more than 1,500 miles (2,400 kilometers) long, 43 miles (69 km) wide on average, and just about 7 miles (11 km) deep at its lowest, a spot that's southwest of Guam and is known as Challenger Deep. That depth measurement wasn't easy to get, and further information is, at the time of this writing, still forthcoming because of the difficulty getting to the bottom.

Though the Mariana Trench was first surveyed in 1951, the first time a human dipped to the bottom of the trench was in November 1960, when U.S. naval officer Don Walsh (1931–2023) went down in a bathyscaphe with Jacques Piccard (1922–2008) more than 35,000 feet (56,000 kilometers) into the Challenger Deep; the machine they used was designed by Piccard's father, Auguste (1884–1962). The second time the Challenger Deep was explored by a human happened more than half a century later, when filmmaker James Cameron (1954–) went down in a submersible vehicle in 2012. On that trip, Cameron also set a world record for a solo descent into the ocean when he descended nearly 36,000 feet (57,900 kilometers) underwater.

Since then, other teams have visited the Mariana Trench.

The Mariana Trench is one place on Earth that humans probably can't easily live (for now), because of the water pressure and overall temperature. The

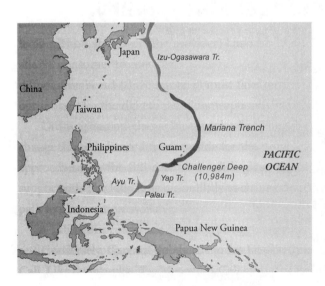

The Mariana Trench is a crescent shaped chasm at the bottom of the Pacific Ocean. It is about 1,580 miles (2,550 km) long and 43 miles (69 km) wide.

> **FAST FACT**
>
> Despite the overall cold temperature at the bottom of the Mariana Trench, you can still cook an egg there, theoretically: throughout the bottom, scientists have found thermal vents from which water flows at up to 700°F (370°C). This water comes from Earth's core and releases minerals that sustain the life so far down.

surprise is that other creatures find the floor of the Mariana Trench a fine place to hang out.

The 1960 expedition found an "ooze" at the bottom of the Mariana Trench, and a few fish that may have been misidentified then. Mud samples contained tiny organisms that thrived in the soil there. Subsequent visits via camera confirmed larger non-fish creatures, fish of varying size, a handful of species that scientists had never seen, and at least one creature that couldn't be identified. The dumbo octopus lives in the Mariana Trench, as do the goblin shark and several types of snailfish.

Notable Names: Breaking Barriers in Space

Born in Hampton, Virginia, Mary Jackson (1921–2005) grew up during a time when Jim Crow laws were enforced, which meant that career options for young African Americans were few, and fewer if one was female as well. Nonetheless, when Mary was old enough to attend college, she went to Hampton Institute and graduated in 1942 with a double degree in physical sciences and math. She almost immediately landed a job in a black school in Maryland and set to work with young students, which was one of her lifelong passions.

After a year teaching, Mary returned to Hampton to work on the home front in World War II. She married, started a family, and had a series of non-career-related jobs before landing at the Langley Memorial Aeronautical Laboratory's all-black West Area Computing Section in 1951, working in the section's pool of general computer operators.

The department was strictly segregated. Black employees could not even use the restrooms set aside for white employees.

Two years of that, and a change came to Mary in the form of an offer to move over to the Supersonic Pressure Tunnel to work with engineer Kazimierz Czarnecki (1916–2005) to perform experiments working with wind tunnels. Czarnecki was pleased with Mary's work, and he encouraged

Mary Jackson, shown here in 1977 at the Langley Research Center.

FAST FACT

The year after Mary Jackson accepted a job working in a segregated computer department, the U.S. Supreme Court took on the court case *Brown v. Board of Education*. It was settled two years later, in 1954, when justices unanimously ruled that separating children by race in public schools was unconstitutional.

her to reach for more education and become an engineer—but there was a catch: Mary would have to apply for special permission to study with white students in an all-white building. That didn't bother Mary; she garnered the permission, took the classes, passed the exams, and in 1958 became NASA's first black female engineer.

For 20 years, Jackson worked as an aerospace engineer with a focus on the winds surrounding an airborne plane. After rising about as far as she could (given the glass ceiling limitations on women in the workplace at the time), she left engineering in 1979 and took a job as Langley's Federal Women's Program Manager.

There she tapped into her passion to help women on their way up the career ladder at NASA. She retired in 1985, but she and her husband never stopped offering welcome assistance, both socially and professionally, to new Langley employees when they arrived in town.

Ancient World: Would You Survive an Attack by a Dinosaur?

Years ago, before we became enlightened on basic timelines, it was common to see movies and cartoons in which "cavemen" and dinosaurs lived side by side. Now we know that dinosaurs and humans were (mostly) not alive at the same time. Still, what would happen if they were? Could *you* survive a dinosaur attack?

To answer that, you should first remember that dinosaurs never had experience with humans, so you'd be an anomaly to them. Even if you suddenly found yourself standing before a gigantic creature with serious teeth, you're probably not going to be perceived as prey for at least a few seconds to a few minutes. Use this to your advantage, if you ever fall into a time warp and get unlucky; if you're quiet and stealthy, you might have time to *quietly* sneak away.

Back to the question of what would happen: there is, of course, the issue of size. Getting attacked by a pterosaur probably wouldn't be much more dangerous than if you were pecked by a flock of chickens. The bottom line is that most little dinos would be no problem for you, assuming that you had modern brainpower and good running shoes. You might be able to pick them up and cast them aside. Even better, they may be as wary of you as you would be of them.

Beware, though: some little dinos may have sharp talons. Ugh.

Next, we'd have to consider whether the dinosaur in question is a herbivore, like a stegosaurus or a triceratops. Since you wouldn't resemble a plant, those big guys wouldn't look at you twice, other than maybe in curiosity. Your biggest danger from them would likely be that they might stomp on you or whack you with a tail if you weren't quick in getting out of the way. That would be especially true if you were near a nervous group of *Gallimimus*, each of which is thought to be able to run at about 40 miles an hour.

Nope, let's be serious. When it comes to danger, we're talking something big and fast.

A Tyrannosaurus rex comes to mind first, right? But the fastest dino would be the Deinonychus, which lived some 125 million years ago and which scientists believe could run at a speed of up to 70 miles (about 110 kilometers) per hour. Still, the Deinonychus wasn't very tall—not quite the height of an average adult human, but weighing around the same. And he was a carnivore, so there's that.

But anyway, for the sake of argument, let's say that the dinosaur that's got his eye on you is the T-Rex because you know who he is. You've seen his movies.

The bad news is that he's big.

Also, the good news is that he's big—that makes him slow on the uptake. Depending on how close you were standing to the beast when he noticed you, if you were just out of reach when he gnashed his teeth at you, you're probably going to be fine. The truth is, you're likely just a hair faster than a T-Rex, giving you a pretty good chance of scootching past his enormous head and his gaping maw. And if you're fleet of feet, well, that's even better because a T-Rex couldn't run. Scientifically speaking, his muscle mass and leg bones simply weren't capable of it.

But look, let's say that you're a total doofus and you wait to see what happens here.

You won't like it. The front legs of a Tyrannosaurus rex are notoriously small, but those

Scientists believe that as many as 2.5 *billion* T-Rex dinosaurs have existed from the beginning of their time on Earth to the end of their reign. They also believe it's possible that T-Rexes hunted in packs.

In the 1960s comedy cartoon *The Flintstones*, wife Wilma Flintstone often serves up bronto-meat for dinner. A brontosaurus is the same creature Fred Flintstone slides down in the show's opening.

aren't your concern. What is, are his teeth, which will crush you with a force of more than 430,000 pounds (more than 195,000 kilograms) per square inch. Not fun at all.

Plants and Animals: Night-Night, Sleep Tight

Whatever you do, don't read this entry before going to bed. There's a psychological process called "suggestion" that happens when you hear or read about something and subsume it in your own life. You've been warned.

It starts out with a tiny tickle. And then another one. Soon you might feel as though you're lying on a pad of wool. You get out of bed and twist around to look at your side, or the underside of your arm, or your lower leg, and you see small red bumps, often three in a row. Uh-oh. Hate to say it, but you've been hanging out with bedbugs.

While the fact is that there are nearly a hundred different species of bedbugs in the world, it may come as the tiniest bit of relief to know that just three of them are known to dine on humans. Small bugs, small comfort, right? Well, how 'bout this: most

Bed bugs have become an increasing problem in American hotels in recent years, thanks to increased tourism and other foreign visitors.

of the problems in the United States come from just one kind of bedbug, *C. lectularius*—although it's a fact that *C. hemipetrus* has crawled into Florida and California lately. Those are the bedbugs that like to take a chunk out of humans, but they'll bite other mammals and birds if they miss you for dinner.

Bedbugs have plagued human sleeping spots for about as long as humans have had beds. Scientists believe that bedbugs were originally *bat*bugs, but that they adapted to humans living in caves. Ancient Greeks and Romans were well aware of the pests, as were Chinese and Egyptian sleepers. Bedbugs are equal opportunity pests, in that they plagued commoners and royalty alike in ancient times.

They're also notorious hitchhikers. Then, they came through trade routes and merchandise; today, they come into a hotel or motel by catching a ride in someone else's suitcase, backpack, jacket, or briefcase, and they'll stay in the room long after that person has checked out. As soon as you check in, they climb aboard your belongings, follow you home, and move in with you for a spell.

Experts think that the boom of bedbugs over the past few decades is, in fact, caused by humans and our love of travel. In addition, the little buggers have become resistant to insecticides.

Since knowing what you're dealing with is the first defense in dealing with bedbugs, it helps if you know what they look like.

Adult bedbugs are small—no more than a quarter inch (a little more than a centimeter) long—but not so small that you can't see them with the naked eye. They're reddish-brown and about the size of an apple seed; that's what they somewhat resemble too. The nymphs (juvenile bugs) are small and very hard to spot; the whitish to yellowish eggs are tiny but visible if you look. If the infestation is bad enough, you can smell bedbugs, but don't let your nose be the judge. Look for reddish residue in mattress creases or smeared fecal matter on the side

> **FAST FACT**
>
> The hyphenated word "willy-nilly," which can abso-lutely be used to describe the movement of a bug, comes from an archaic phrase, "will I, nill I," which means that the speaker is both willing and not will-ing, or is waffling on a decision. Eventually, the word came to mean "at random" or "all over the place."

of the bed; bedbugs can hide in or on both uphol-stered and wood furniture, pant leg hems, curtains, the backs of framed pictures, woodwork joints, the-ater seats, and carpet. You may also see egg cases or live bugs. Note that bites are not reliable indicators of bedbugs, since bedbug bites are usually accom-panied first by a form of anesthetic so you don't nec-essarily feel them right away, and other kinds of bugs bite too. Generally speaking, though, bedbug bites appear in a row of three or more itchy bumps.

Small as they are, bedbugs will travel across a room to get a meal, and they'll dine on your pet or a wild bird just as easily as they dine on you. They seem to prefer feeding at night, but will eat any time a meal presents itself. And they often scoff at the weather: bedbugs can survive and feed in temps as low as 45°F (7°C) and as high as 113°F (45°C). Tropical breeds even laugh at *that* temperature.

Fortunately, there's no evidence that bedbugs can carry or transmit viruses or bacteria to humans. They are just icky, that's all.

Human Body: Mental Health, Eugenics, IQ Tests, and Some Breakfast Cereal

If you were the parent of a developmentally disabled child in the early part of the twentieth century, the options you had for your child were often very limited. Most recommendations were that parents send their child away to live in an "institution," forget the child ever existed, and go on with their lives as if nothing happened. There was a certain amount of shame in having such a child and acknowledging the possibility of genetic developmental disabilities in the family.

Heartbroken parents might have felt lucky to find a place for their children then, away from society. They may have welcomed the presence of a place like the Walter E. Fernald Developmental Center near Boston.

Founded by Samuel Gridley Howe (1801–1876) as the Massachusetts School for the Feeble-Minded in 1848 with a $2,500 grant from the State Legislature, it was located in South Boston and was meant to help mentally challenged youth (mostly boys) to get an education in a simple trade so that they could grow up to be social, responsible, and productive adults. Howe, a reformer and abolitionist, had experience with starting such institutions: he was also the founder of the famous Perkins School for the Blind.

In 1888, the state approved a move for the school to a former farm in Waltham. This allowed for expansion, and the Massachusetts School for the Feeble-Minded proved to be a success, and officials

Abolitionist and physician Samuel Gridley Howe helped found schools for the blind and the mentally challenged with the intent of educating and training young people to be productive citizens.

pushed to offer help to adults with disabilities. What's more, after he took the reins, its third superintendent, Walter E. Fernald (1859–1924), gave the school a new, then-cutting-edge scientific focus.

He and other officials at the school around the turn of the last century hoped to focus on eugenics.

Based loosely on misunderstood Darwinian theories and other societal ideas of the day, the "science" of eugenics was first developed by Sir Francis Galton (1822–1911), who stated that people of color, Jewish people, Roma, developmentally disabled individuals, and the poor were doing great harm to the gene pool in general. The obvious solution was to manipulate how those humans reproduced in a given society; if they can't have babies, eventually, those "undesirables" could be all but eliminated. Despite that this is a lot of bunk, many celebrities and big thinkers of the late 1800s and early 1900s bought into the idea of eugenics and thought it was a valid goal. If it sounds a lot like racism, though, that's because it is.

And so a perfect storm began brewing: In 1905 a new kind of fancy test became available, making it easy to separate people even further. Named after its creators, Alfred Binet (1857–1911) and Theodore Simon (1873–1961), the Binet-Simon IQ test meant that anyone could be tested and labeled as intelligent or not. Surely there were a good number of adults who took the test to boast, but the harm came from elsewhere: children who tested below the normal mark for whatever reason were often removed from their families and institutionalized.

The entire concept of eugenics was debunked at the end of World War II, after the Nazis killed millions of people they deemed undesirable while also working to breed a superior race.

To further the storm, Fernald had a seat on the board of the Eugenics Society, and he was surely aware of the Binet-Simon IQ test. You have to wonder how he saw this test in relation to the school he was running. If he liked it, well, Fernald wasn't alone.

Suddenly it seemed that everybody was doing (or having done to them) the IQ test; by 1913, the test was being given to new immigrants at Ellis Island. Institutions sprang up all over the United States to accommodate the sudden influx of patients who were likely *just fine* before being tested but, for whatever reason, tested poorly. That included, and was especially true of, children from poor families and those who were orphaned or were simply unwanted and were dumped off. Many of those kids were of normal intelligence and fully capable of learning, but they often never got the chance—children at the Massachusetts School for the Feeble-Minded weren't formally educated.

Children who went to live at the school and were able to work were expected to do some sort of task and often toiled as free labor to run the place; there was even a "ratio" of able to non-able that authorities recommended the school keep. Those kids who tried to run away—and a surprising number did—knew that if they were caught, terrible punishment awaited them. Reports exist that say abuse from staff was common. To say that there was little to no oversight in this situation is to put it mildly.

Fernald died in 1924, but he left a legacy behind. Early in his life, he'd advocated forced sterilization of school patients who tested on the lower end of the Binet-Simon IQ test, but he later reversed his opinion. Instead, he decided that strict segregation was better.

A year after his death, the school was renamed in his honor.

At some point in the mid-1940s, things kicked up a bit.

There doesn't seem to be a record for who thought of it, but from 1946 to 1953 dozens of male children were invited to join a "science club" with incentives to belong, such as trips and food treats. Once in this "club," they were also given small doses of radiation mixed with their breakfast cereal, or injections of radioactive calcium. The aim of this whole years-long episode, which was cosponsored by Harvard, MIT, and Quaker Oats, was to study the effects of radiation on children who had no say in the matter. The boys' guardians reportedly gave consent for this, but most agree today that the consent forms weren't exactly informative, clear, or well explained. It's been said that the amounts of radioactive substances that the boys received were small and probably harmless.

Probably.

By the 1970s, society had changed its mind about the idea of institutionalizing developmentally disabled people, and Congress ensured that places like the Fernald School would be run more humanely. In 1994, Senate hearings confirmed the abuse that so many children had endured and the experiments to which many had been subjected. Parts of the original Massachusetts Home for the Feeble-Minded were entered into the National Register of Historic Places that same year. A class-action suit was settled in early 1998 for the radiation inci-

Established in 1888, the Walter E. Fernald Developmental Center in Waltham, Massachusetts, is the oldest publicly funded school for the developmentally disabled in the United States.

FAST FACT

The Nobel Prize in Physics in 1901 went to Wilhelm Röntgen (1845–1923), who, in November 1895, produced and discovered what we call X-rays or Röntgen rays. In 2004, the International Union of Pure and Applied Chemistry named element 111 *roentgenium* in his honor. The unit of measure *roentgen* was already named after him.

dents, and some former Fernhold residents were later compensated for their suffering.

In November 2014, the last patient was moved to a home in the Waltham, Massachusetts, community. The City of Waltham owns the site of the former school.

Human Body: Achoo! God Bless You!

Your nose feels funny and you know what's coming. You take a deep breath—ah. Ah. Ah. Your nose wiggles and CHOOOOO!

Gesundheit. So why did you just sneeze?

The basic reason is that something entered your nose that irritated the mucous membranes inside it. That's the simplest explanation for sternutations, but there's more to a sneeze than meets the, um, nose. Ah.

It starts with that foreign bit of something, whether an allergen or a trigger that tells your nose that something's wrong and that something needs to *go, now*. That information goes to your brain's sneeze center (yes, that's a thing), and the sneeze center notifies your diaphragm, chest muscles, abdominal muscles, and vocal cords to get ready to forcefully, definitely expel the problem. Sneezing, overall, keeps your body safe. Ahhh.

Read that again: forcefully. The sneeze droplets—possibly as many as 100,000 of them—come out of your nose and mouth at up to 100 miles (more than 160 kilometers) an hour and can land some 6 feet (2 meters) away from your face. If those droplets land anywhere someone might touch, that someone could get sick, so it's important to know how to sneeze correctly, which is to sneeze into a tissue or into your elbow so you don't spread germs on your hands. Oh, and don't try to hold in a sneeze; it could harm your sinus passages. Ahhhhh.

Your heart doesn't skip a beat when you sneeze, but a big sneeze may cause a delay before the heart resumes its regular thumpa.

You can also try to keep your eyes open when you sneeze, but good luck. Scientists aren't sure why, but it seems to be a reflex that your eyes close when you sneeze. Maybe it's nature's way of keeping the irritant out of your eyes, but that's just a guess. Ahhhhh....

Dogs, cats, elephants, even birds and reptiles can sneeze. Iguanas are said to be first-class sneezers, and they use sneezes to clear their bodies of digestive material. But sharks? Nope, sharks can't sneeze.

As for frequency of sneezes, well, that's going to depend on the individual and any allergies they may suffer. Studies show that 95 percent of the people without allergies sneeze and nose-blow no more than four times a day, on average. Any more than that, on a regular basis, and it's probably a good time to check with a doctor. Ahhhhh....

For some unknown reason, it is pretty much impossible to keep your eyes open when you sneeze. Next time you feel a sneeze coming on, try to keep your eyes open and find out if it's true.

You can sneeze in your sleep, but only in the very earliest sleep-cycle. For reasons that are not clear yet, sleep suppresses the urge to sneeze, even if an irritant is introduced directly into the nose.

Just know that allergies and irritants aren't the only reasons you sneeze. Bright light can make some people sneeze, and that's an inherited thing. Plucking one's eyebrows can irritate a nerve in the general area of your nose and cause a sneeze. Sex can make you sneeze too, and so can working out in the gym. Ahhhhh …

CHOO! While being famous is often fun, be glad you're not a record-breaker this time: the world record for sneezing went to a woman in the U.K. who suffered a sneezing fit that lasted nearly three years. Bless you!

FAST FACT

In the same year (1983) that Billie Jean King made it to Wimbledon for the fourteenth time in women's tennis, Ivan Lendl made it to Wimbledon on the men's side for the first time. Getting there must've been an extra challenge, since British tennis is sometimes played on grass courts, and Lendl is highly allergic to grass.

Plants and Animals: Oyster Spit

Audrey Hepburn really knew how to do it. Jackie Onassis had the knack, for sure. Michelle Obama and Sarah Jessica Parker are old hands at it. But it doesn't take a special kind of person to wear pearls well; anyone can appreciate the elegance of a pearl necklace. It might help to know how a beautiful, perfect pearl came to be.

To know how a pearl is made, you must start with an oyster.

For at least 15 million years, oysters—a word to describe several species of saltwater bivalve mollusks—have populated brackish, salty shorelines along rocks or reefs or, later, wooden piers or have created their own reefs. Their lives are easy and uncomplicated: basically, oyster larvae are born and then float for two or three weeks, whereupon they will attach to almost anything, including industrial or man-made beds, for a place to grow and live. At that time, they're known as spat. Once spat are attached, they mostly spend the rest of their lives in one spot.

Long before they get to where they're going—about 12 hours after birth, to be exact—spat begin taking calcium from the water, secreting proteins and minerals from their own bodies, and layering the mixture on themselves. Over time this creates

Natural pearls are formed when the oyster covers an irritant with nacre (calcium carbonate) to protect itself. Real pearls from an oyster are generally not round.

the shell of the animal, a shell it'll have until it dies, because oysters don't shed their shells. They add to them as they grow. It takes an oyster about three or four years to become fully mature.

And the whole time it's doing that, the oyster is cleaning water while it eats, since an oyster is a filter-feeder. That means it takes through its gills large amounts of water containing plankton, algae, and things oysters can eat, along with pollutants and other detritus. Whatever isn't food is pooped out as waste called pseudofeces, which settles on the bottom of the bay. To help you understand how important this is, a single oyster can filter and clean up to 50 gallons (189 liters) of ocean water each day.

If we're talking oysters after a meal, they're still useful. Near the water, oyster shells can be recycled to make more oyster reefs. Away from the coast, you can recycle oyster shells by crushing them and using them as a soil additive in your garden.

Here's where the pearl comes in.

You know what happens when you're eating, and a bite of food goes down "the wrong pipe"? That happens to oysters: a small amount of food or any kind of small foreign body can get trapped inside the animal's shell where it very much doesn't belong. The oyster immediately knows that the irritant is there, and to protect itself, it quickly coats the food with aragonite and conchiolin, two substances that also make up an oyster's shell and form nacre, which you may know as mother-of-pearl.

Over time, the creature continues to secrete these two materials around the irritant, layer upon layer upon layer, more and more and more, until a pearl is formed. If you have a

> Mother-of-pearl doesn't rate super high on the Mohs hardness scale— it rates somewhere between 2.5 and 4.5, which is between a fingernail and a knife blade.

> Freediving is a sport in which participants take one breath, dive into water, and go as low as they can physically withstand before needing to come up for air. Freediving is an ancient activity; it's been around for about 9,000 years and was once used to harvest sea sponges and pearls.

natural pearl, you can actually see the layers in very bright light.

Natural pearls, the ones that are most highly prized because of their rarity, occur in nature without any interference from humans; cultured pearls, on the other hand, require a human to insert the irritant into the oyster's shell to jump-start the process. Still, as with natural pearls, cultivating a pearl offers no guarantee of size, color, condition, or even the end existence of a pearl. In both cases—cultured or natural—a pearl can form in as little as six months, or it can take years.

Early Europeans were said to have been wild about pearls, and they sent out harvesters who captured oysters almost to extinction, but in an ugly way: oysters were found, opened, and thrown away whether their shells contained pearls or not. By the early 1900s, Kokichi Mikimoto (1858–1954), a Japanese entrepreneur, had figured out a way to insert an irritant into an oyster shell without killing the creature within.

Freediving is a sport in which participants take one breath, dive into water, and go as low as they can physically withstand before needing to come up for air. Freediving is an ancient activity; it's been around for about 9,000 years and was once used to harvest sea sponges and pearls.

Technology: Anyone Find That Bomb I Lost?

At some point in your life, you learned (or will learn) that you have to keep track of your belongings, or you won't have them anymore. Find the Hidden Item is a basic preschool game. So how did the United States come to lose several nuclear bombs?

In the early years of the Cold War, well after World War II was over, the United States continued to fly airplanes over Europe that were loaded with thermonuclear bombs. At least 32 times, in what the government calls "Broken Arrow" incidents, there were serious accidents involving contamination, accidental launchings or detonation, or total loss of warheads. In six of those Broken Arrow incidents, the United States lost nuclear weapons that couldn't be found or retrieved somehow. Fortunately, five of them were incapable of detonation and were relatively safe, all things considered.

Do the math. That leaves one.

Here's what we know:

- *In March 1956*, a B-47 Stratojet took off from Florida and was heading for an air force base overseas, loaded with two nuclear weapons. The first refueling was normal; the second was supposed to take place over the Mediterranean Sea, but the plane never arrived for this second, in-flight, refueling. Officials searched and searched, but they eventually called off recovery efforts. Two cases of nuclear material were aboard, protected in cases, so the chances of a nuclear detonation were zilch.

- *In July 1957*, a C-124 cargo plane took off from Delaware with two nuclear bombs and one nuclear capsule when the plane experienced a loss of power on the first two engines.

A B-47 Strato-jet like the one pictured disappeared in 1956. It was carrying two nuclear warheads. Fortunately, the bombs were inoperable and had no chance of detonating.

Because the weapons were not fully operational and ready for detonation, all were deemed safe, and two were jettisoned somewhere over the Atlantic Ocean. A search was conducted, but no debris was found.

In February 1958, a B-47 bomber with a weapon onboard left Florida on a training mission; the aircraft was near Savannah, Georgia, where it accidentally collided with an F-86 fighter plane. The B-47 struggled to land at Hunter AFB in Georgia, but it could not, and so, to avoid an explosion at Hunter AFB, the weapon (without its nuclear capsule) was released near Tybee Island but was never recovered. Its cargo—mostly uranium-235—was not deemed dangerous, since it could not detonate by itself.

In January 1961, a B-52 carrying two bombs broke apart because of structural damage, and both nuclear bombs that were aboard were accidentally released above North Carolina before the plane completely fell apart. Five of the eight crew survived and notified the military, who went to retrieve their equipment. One of the bombs was found hanging from a tree, supposedly with its nose just touching the ground. The second one had fallen apart nearby. One weapon contained uranium. Both were one step away from detonation.

In December 1965, a Navy A-4 Skyhawk newly loaded with a thermonuclear bomb rolled off the side of the USS *Ticonderoga* while it was being moved, with the pilot and warhead both inside. The plane quickly sank into such a depth that retrieval was impossible, and to this day no one knows

FAST FACT

More than 300 dogs live in and around the Chernobyl nuclear reactor site in Ukraine, the place where a nuclear reactor melted down in 1986. While rabies is a concern, scientists are more interested in the long-term effects the disaster and the radiation fallout have had on the pet dogs. Specifically, they're studying the canines' DNA to see what it might take to survive a nuclear disaster.

if the bomb could or did detonate at that low pressure point in the ocean. The pilot was lost.

In January 1966, a midair accident between a fueling B-52 and a KC135 refueling tanker resulted in the accidental release of four nuclear bombs over Palomares, Spain. One landed safely, two detonated and scattered plutonium over about 2 miles (3 kilometers) of land, and the fourth one went into the ocean. The Navy retrieved it a few weeks later; the U.S. government cleaned up around 1,400 tons (roughly 1,423 metric tons) of contaminated soil and deposited it in at an approved storage site.

In January 1968, a B-52 coming from New York crashed some 7 miles (11 kilometers) from Thule AB in Greenland. Four nuclear bombs were accidentally released over Greenland. Most of the crew survived, but four nuclear weapons aboard the bomber burned in the plane's fire, which resulted in some nuclear contamination. The U.S. government removed some 237,000 cubic feet (6,700 cubic meters) of contaminated ice, snow, water, and plane debris in the subsequent four months, and took it to an approved site.

In May 1968, the USS *Scorpion*, a nuclear-powered submarine that was returning from a months-long training mission, was lost some 450 miles (720 kilometers) southwest of Portugal under nearly 10,000 feet (3 kilometers) of water. Rumors that Russia was somehow involved were proved to be just that—rumors. Ninety-nine crew members were lost. The Scorpion was known to have been carrying some sort of nu-

In 1968, a B-52 carrying four nuclear warheads crashed just 11 miles away from the Thule AB and leaked radiation into the surrounding area.

clear weapons, but not much else is confirmed, and most details remain classified.

These are the things we know because they were just declassified in 1980, so there may be more information to learn in years to come.

And while all this sounds absolutely dreadful, don't relax yet. It's not clear how many nuclear bombs the Soviet Union lost, or where. The assumption is that they lost theirs in the same manner as did the United States—in the ocean, safely beneath the water—but nobody knows for sure.

Never underestimate the power of wind and water: When Hurricane Wilma hit Florida in 2005, millions of Floridians were left without power when two nuclear power plants in the area were shut down in anticipation of the storm's damage. Nuclear power plants are built to hold up to most disasters, including hurricane-force winds, but officials often decide to err on the side of caution. A few days without power is nothing compared to a possible nuclear plant disaster.

Plants and Animals: Don't Pet a Rabid Raccoon and Other Zoonotic Diseases

Good old Fluffy and Fido. They're your best friends, housemates, members of the family, your best sounding-boards. Can't imagine life without them, can you? You share your meals and your bed with your beloved pets. So, what might they share in exchange?

At the top of any list of zoonotic diseases (defined as any disease that can jump from animal to human) is rabies, perhaps because its side effects are so very frightening. Rabies is a viral disease spread through the saliva of an animal that is affected, as well as through tissue, tears, and nasal secretions—

which means transmission is most often through a bite but, though rare, can occur via any break in the skin through which the secretions of a rabid animal can pass. Domestic dogs are responsible for up to 99 percent of all rabies transmission cases.

When the rabies virus is passed to its next victim, it ultimately affects the nervous system; in humans, it causes hyperactivity, excitability, fear of fresh air, agitation, confusion, paralysis, and possibly pain upon swallowing—and thus a fear of water (which is where its original name, hydrophobia, comes from). Once those symptoms appear, death occurs in nearly 100 percent of all rabies victims, even despite treatment. Fortunately, there is a vaccine for rabies.

Rabies, by the way, is found on every continent except Antarctica.

Yep, rabies is why you get your pets vaccinated and you don't mess with wild animals for fun.

But Kitty can get you in several ways.

There is, of course, rabies, which cats can get if they're allowed to hunt outside. Then there's cat scratch fever, which is not just a rock & roll song— it's a real bacterial infection caused by the *Bartonella henselae* bacteria carried in a cat's saliva or claws, and it causes mild fever and swelling in the lymph nodes near the affected wound. Indeed, most

Rabies is a scary and deadly disease! It is very important to get your pets vaccinated!

Rabies is not transmitted through a rabid animal's urine, fecal matter, or blood.

people won't get too sick from cat scratch fever, but immuno-compromised folks can be affected neurologically.

Thought to be one of the country's most common parasitic infections, toxoplasmosis can come from uncooked or contaminated meat, but most people know it by its association with cat feces. Cleaning a litter box can put you in proximity of the toxoplasma parasite, as can handling anything that's come in contact with feline fecal matter or soil that cats have defecated in. While it's true that most people's immune systems are strong enough to deal with the parasite—in fact, the CDC says that 40 million Americans carry the parasite with no ill effects—newly pregnant and about-to-be-pregnant women and those with bruised immune systems need to be especially cautious of toxoplasma.

And then there are allergies, which can come from cat saliva when they clean themselves and their fur. Allergies can cause respiratory symptoms and make you miserable.

Snakes, reptiles, and turtles are not entirely safe. From them, you can get *E. coli*, salmonella, and campylobactor, all diseases of the human digestive system that can ultimately cause death if not treated. Baby ducks and chicks can also cause salmonella; and so can pet birds such as parrots and parakeets, which can also cause psittacosis, resulting in a flu-like illness.

Just touching animals can make you sick, and so can ingesting them. Eating undercooked or raw meat from an animal that has been infected by the trichinella parasite (usually wild game or pork) could cause trichinosis (also called trichinellosis). You may feel as if you have the flu at the beginning of a trichinosis bout, but it gets worse. Let the disease fester, and you'll have heart and breathing difficulties. Don't seek medical attention, and you could die.

Roundworms aren't a parasite that just your pet gets; people can get them too, especially if they eat undercooked meats.

As for wild animals, bats and birds can pass on the fungus spores of histoplasmosis, which causes problems with breathing. Roundworm can be found in the feces of raccoons and dogs, among other animals; another unsavory worm is hookworm. Ringworm is not a worm at all but a fungus that causes a ring-shaped patch that pets and humans can pass to each other.

So what can you do to make sure you and your family don't get sick? Experts recommend one big thing to protect yourself: wash those hands!

FAST FACT

Dr. Anna Wessels Williams (1863–1954) spent much of her career studying bacteria in the field of bacteriology. In 1896, she traveled to the Pasteur Institute to look for a toxin that she could use in the development of a vaccine for scarlet fever; she had done something similar for diphtheria. While in Paris, Williams discovered a culture that she ultimately developed into a rabies vaccine. She was also instrumental in creating a better, more efficient way to test the brain matter of a creature suspected to have rabies to determine if treatment was warranted or not.

Human Body: More Poo, Pee, Blood, and Eww

- *The mucus in* your nose (can we just call it "snot"?) is there to protect your lungs by keeping any smoke, dust, or bacteria that you might breathe in from harming you. The bad stuff is collected, and it slips into the back of your throat, diverting to your stomach where it will be rendered harmless. Snot, by the way, is at least 95 percent water, and you make almost a half-gallon of it every day.

- *Sanitized human urine* makes a good fertilizer, according to several studies.

- *That stuff that* dries up and collects in the corners of your eyes every night is called rheum.

- *The average human* defecates seven to eight times a week and passes gas up to 15 times a day.

- *There are four* main blood groups—O, A, AB, and B—but because your blood can be either RhD (rhesus factor D) positive or RhD negative, there are eight blood types altogether. The rarest of those eight is AB negative, which a mere one percent of the world's population has.

- *Unless you suffer* from hemophilia or any other sort of non-clotting disorder, doctors say that it's nearly impossible to bleed out from a paper cut. It *is* possible, according to those in the know, to create a sharp knife from paper by folding it a certain way.

- *Human eyes shed* three kinds of tears: basal tears, which lubricate and protect your peepers; reflexive tears, which do about the same thing, especially when you get dust or dander in your eye; and psychic tears, which are shed due to strong emotions. Scientists say that the latter can help your body get rid of stress, so go ahead and cry it out.

- *The wax in* your ears can tell an anthropologist where your ancestors lived. People of European and African descent generally have "wet" earwax that is yellow in color. People of East Asian descent generally have flaky earwax that is

> **FAST FACT**
>
> Even recently, some people believed that leaving plants in a sickroom or hospital room depleted the oxygen and harmed the recovering patient there. That's a myth. In a report published in 1989, NASA scientist Bill Wolverton (1932–) proved that "a closed ecological life support system," including certain kinds of houseplants, will completely clean the air in a 45-feet by 16-feet Biohome and make it safe for its inhabitants.

white or gray. Earwax, by the way, migrates from inside your ear to the outside, taking debris and bacteria away from the eardrum and sweeping it out.

Earth Science: Can You Dig It?

So, you say you'd love to be a scientist someday? You've had dreams of digging in the dirt for dinosaurs or other fossils. You want to know what it's like to pull valuable, interesting rocks from the soil. Geology, earth science, and paleontology fascinate you.

And you're in luck.

There are many places around America that are open and ready to satisfy your curiosity and your dreams of digging. Put on your coat, grab your trowel, double-check the open times and rules of these spots, and let's go.

Starting on the eastern side of the United States and working our way west, the Crystal Grove Diamond Mine and Campground in St. Johnsonville, New York, is where you'll find gorgeous clear quartz crystals called Herkimer diamonds. Bring your own tools or rent them if you want. Know that you may need a pick because busting rocks could be worth the work involved.

The Penn Dixie Fossil Park & Nature Reserve in Blasdell, New York, charges a small admission fee, but your chances of finding a Devonian Period fossil here are said to be pretty high. Check the reserve's website; they're not open every day. You can also buy admission tickets online.

If you're truly serious about paleontology, you need to visit Big Brook Preserve in Colt's Neck, New Jersey. Ancient shark's teeth and other cool things can be found in the dirt surrounding local waterways, so bring waders along with your regular tools. Before you head out, be sure to check the website first; Big Brook has lots of rules you must follow and a limit on the number of fossils you can keep per day.

The Montour Preserve Fossil Pit in Danville, Pennsylvania, is basically a one-acre pit of shale, in which plenty of fossils are waiting to be found. Bring your own tools; it's recommended that you wear long, sturdy pants because shale edges can be sharp. Fossil Park in Sylvania, Ohio, is a hop and a skip from Danville, Pennsylvania, and it's full of shale that's trucked in and dumped for your digging pleasure. There's no charge for this park, but there is a fee for a guide, if you wish. Bring your own tools.

Heading a little southeast, Emerald Hollow Mine in Hiddenite, North Carolina, is the country's only emerald mine that's open to public digging. There are sluiceways to sort through, or you can dig in the dirt to your heart's content for a little bit of a fee; if you're lucky, you'll come home with an emerald, tourmaline, aquamarine, topaz, or garnet find.

Also in North Carolina, check out Gem Mountain Gemstone Mine in Spruce Pine. Dig there and you might find aquamarines, moonstones, and rubies—and when you're about ready to head home, take your finds over to an on-site appraiser who'll tell you what you have.

One more in North Carolina: check out the Cherokee Ruby and Sapphire Mine in Franklin, which is normally open from April to October—but check their website.

Perhaps the most famous of them all, Crater of Diamonds State Park in Murfreesboro, Arkansas, is America's only open mine where you can find diamonds and keep what you find. Bring your own trowel or rent one of the machines for deeper digging; they too offer an on-site appraiser so you don't take home random, worthless dirt.

Though it's a little rough, with no running water or shelter from the elements, Mineral Wells Fossil Park in Mineral Wells, Texas, is a great place to find fossils of sea creatures that lived millions of years ago. Bring lots of water and your own tools and keep what you find.

Also bring your own tools and buckets when you visit Rainbow Ridge Opal Mine in Virgin Valley, Nevada. Open from May to September, this mine has yielded a high number of valuable stones, so good luck there! While you're nearby, take a trip to Bonanza Opal Mine in Denio, Nevada.

An hourly fee is all it takes for you to find fossils at U-Dig Fossils in Delta, Utah. Tools are provided, and there's a good chance that you'll find a trilobite

Diamond hunters sift dirt and water through screens in search of gems at Arkansas' Crater of Diamonds State Park.

FAST FACT

William "Billy" Barker (1817–1894) was a British prospector who emigrated to America in the 1840s, leaving his wife and family behind. Ten years after his arrival in New York, Barker was prospecting in California. When he didn't find the fortune he thought he should have, he and his crew headed to British Columbia where, in 1862, Barker struck several pounds of gold, the largest amount in the area.

or five in the shale there. Bring the kiddos; the place is family friendly.

For the amateur who wants to learn before digging in earnest, sign up for "Dig for a Day" at the Wyoming Dinosaur Center and Dig Sites in Thermopolis, Wyoming. When you do, you get a day of digging with an experienced paleontologist, as well as time in a lab so you know what's involved in the job. There's an admission fee for all this, and yes, it's family friendly.

Finally, you'll want to visit Gem Mountain Sapphire Mine in Philipsburg, Montana, where lucky prospectors find sapphires. This mining spot also offers an on-site appraiser and jewelry makers, so you can wear your gemstone home.

Human Body: Are You an Innie or an Outie?

Long before you were born—before your mother even knew you were there—the collection of a few dozen cells that would be you someday was created. Sperm was deposited and migrated to your mother's ova and fertilized it; then one cell became two became four, and so on. At that point, you were called a blastocyst. Those cells that would be you had taken most of a week to grow; within a couple of days, you attached yourself to your mother's uterus, high up, and later, to a placenta that formed. By that point, you had a full-fledged umbilical cord, a conduit through which you shared a blood source with Mom and received nourishment from her. Literally then, you couldn't live without Mom until many more weeks would pass.

Normally, you were attached to your mother until just moments after birth. Once you were fully born, probably within minutes, a doctor or midwife clamped your umbilical cord to lessen any blood loss. The cord was cut, and you were on your own, kid.

After a few days, what was left of your cord dried and fell off.

And you have evidence of all this on your body for anybody to see. Technically, it's called an umbilicus or a navel, but you probably know it as a belly button.

Chances are, your belly button is the first scar you'll ever get—because yes, that's what it is: it's a scar resulting from the separation of your abdomen

FAST FACT

If you cross-sectioned an umbilical cord, you'd generally see one big vein and two arteries (although there are variations). The vein passes oxygen from mother to fetus through the placenta to the liver; the arteries return the depleted blood and any waste back to Mom to process. And by the way, if your Mom was typically an "innie," her belly button probably turned "outie" to make room for you.

and that dried umbilical cord. What it looks like—whether it's small or large, round or oval, horizontal or vertical, an "innie" or an "outie"—has nothing to do with the way the cord was cut. It has *everything* to do with how that scar was formed.

Nowadays, you'll see innies and outies all over TV, movies, and electronic and social media, but showing one's belly button in Western cultures was considered to be indecent (at best) and illegal (at worst) until about the 1950s—and even then, there were censors everywhere. Case in point: famously, actor Barbara Eden (1931–) wasn't supposed to show her navel on the TV show *I Dream of Jeannie,* although Eden admitted in her biography that her navel peeked out occasionally. The first woman to show her navel on television was Yvette Mimieux, who bared her belly in a two-part *Dr. Kildare* episode in 1964.

Asian and Indian cultures didn't have those kinds of taboos.

The average belly button is not very deep, generally just about a quarter of an inch although obesity can extend that depth. No matter how shallow it might be, though, it's pretty dirty in there: there are

FAST FACT

Studies prove that a fetus's DNA crosses the placenta to the mother and becomes a part of her muscle and tissues. Years and years later, a baby's DNA can be detected in its mother's body. Once you're born and don't need it anymore, the part of your umbilical cord that's left inside of you becomes more or less nothing but ligaments. Those veins and arteries that fed you *in utero*, though, become an important part of your circulatory system, even now bringing blood to many of your internal organs.

A recent survey showed that, after ear piercings, belly button piercings are the next most popular with women, with one in three piercings being in the belly.

at least 65 kinds of bacteria that live in your navel (and possibly more!), not to mention the detritus that collects in there because you wear clothing, and that can leave lint. What's more, most belly buttons don't smell very good, and you can get an infection there frighteningly easily. This is important to know for the 90 percent of us who have "innie" belly buttons; as for the other 10 percent of humans who have "outies," you needn't worry about such things. Outie belly buttons, incidentally, happen in cases of small fetal hernias or if errant tissue causes the scar to grow instead of recede.

FAST FACT

In the latter part of the 1860s, horticulturist William Saunders (1822–1900) received a few small trees from Bahia, Brazil, trees that had been grown from cuttings taken by a grower who noticed that the oranges from the original tree offered particularly sweet (and very cute) oranges with human-like belly buttons. Saunders grafted buds from the little trees onto another, larger tree and propagated more trees. Because navel oranges are seedless, then, the oranges you get in the grocery store all came from one mother tree more than 150 years ago.

Fear not, though: if you wish to change your innie or your outie, plastic surgery can give you a cuter navel in less than an hour, generally.

Final Thoughts: Things That Could End Us

And so we come to the end of this book, and it's been fun, hasn't it? Now, just one question remains: how does it all end?

First, let's take a look at a device that might scare the pants off you, once you know all about it: the Doomsday Clock.

In 1947, Hyman Goldsmith (1907–1949), co-editor of the *Bulletin of the Atomic Scientists*, contacted artist Martyl Langsdorf (1917–2013) to create a clocklike design that he could use for the June issue of the *Bulletin*. It would be the publication's first *magazine*, rather than just a newsletter, and Goldsmith wanted something striking. Langsdorf—who was known by her first name professionally—was married to physicist and Manhattan Project scientist Alexander Langsdorf Jr., so she was an easy choice for the assignment.

Martyl created the Doomsday Clock as a metaphor for how close a catastrophic disaster loomed and how long humans would be able to survive the ensuing apocalypse.

The biggest danger to humans in 1947 was the arms race between the United States and the Soviet Union, and the possibility that either side might use nuclear weapons against the other. Post–World War II, people were well aware of the destruction of nuclear weapons, and few had the appetite to see anything like that again. That first year, the clock was set at 7 minutes to midnight—midnight being the End Time. In 1949, when the Soviets actually tested their nuclear arms, the time was

Most predictions of the end of the world involve human folly and the use of nuclear weapons. While the Cold War is over, the potential for a nuclear holocaust remains.

moved to 3 minutes to midnight. The farthest from catastrophe the Doomsday Clock has ever been moved was 17 minutes to midnight, set in 1991. Until 2007, politics and world affairs determined the Clock's time, back and forth, a little here and a little there. That year was when climate factored into a possible coming apocalypse, which added a whole new level of the end of us.

Since 1973, when the *Bulletin*'s editor, Eugene Rabinowitch, died, the publication's Science and Security Board have been responsible for moving and setting the clock. The board comprises nuclear scientists and climatology experts, including several Nobel Prize winners.

In 2020, they set the Doomsday Clock to 100 seconds to midnight.

In 2023, they set the Clock to 90 seconds to midnight.

When you consider everything we humans do to one another and to the planet, there are several ways we could all die in a not-really-so-spectacular exit.

One of the most modern ways could be that the world might end slowly in a climatological apocalypse, with drought in some places, flooding in others, too much heat, possibly boiling oceans, and nowhere for plants to survive to serve as food for us or for other creatures. Some climatologists believe that this end has already been set in motion and that it's going to take a worldwide effort to reverse it.

That's *if* it can be reversed.

Along those same lines, scientists think it's possible that the Sun could die and expand, which would melt Mars and possibly Venus and vaporize humans here on a fast track. Not a pleasant way for us to go, of course, but it would be faster.

Our planet could somehow fall or be pushed out of orbit, or it could be slammed by another planet that has been pushed out of its orbit, or we could be hit with asteroids that are too large to deflect or destroy before they land here. Earth could be a victim of planetary engulfment in which a star or other celestial entity would consume it whole. Again, not fun—especially since we may have some advance notice of this but won't be able to do a thing about it. In that case, all we could do is to watch our own demise.

Evolution—or a lack thereof—could kill us off in the same way it did the Denisovans or the Neanderthals. Would there be another humanoid or humanlike species to replace us? Possibly. Possibly not, depending on a lot of factors that would have to mimic or replicate what happened when the first being came from the ocean to land. Some researchers actually believe that this scenario has already happened once, eons and eons ago.

The first *recorded* End of the World prediction happened in the years 66–70 B.C.E., when Simon bar Giora of the Jewish Essenes said that the Jews' uprising against the Romans would usher in the Messiah for all time.

And then there are the things we'd do to ourselves.

You'd think leaders of powerful nations would know by now what nuclear war would do, but still: nuclear war looms large on the list of Things That Could End Us, as does chemical or biological warfare … which leads us to death by epidemic or pandemic, which has recently become a more plausible possibility.

And then there's the increasingly not-so-1950s, not-so-science-fiction chance that AI and robots could be the death of us all—not directly, but in the hands of someone who might program them nefariously and make them dangerous to other people. For the most part, scientists and robotics experts say that the chances of AI pulling together collectively and rising up are small, but changes in the AI field are made every day.

And so one day, we may look at a robot and think, "This is the end."

Further Reading

Websites

American Lung Association at lung.org

The Center for Disease Control at cdc.gov

The Cleveland Clinic at my.clevelandclinic.org

Discover Magazine at discovermagazine.com

National Aeronautics and Space Administration nasa.gov

National Institutes of Health at nih.gov

National Oceanic and Atmospheric Administration at oceanservice.noaa.gov

The National Park Service at nps.gov

National Public Radio at npr.org

Scientific American at scientificamerican.com

Smithsonian Magazine at smithsonianmag.com

Books and Periodicals

"The 5-Second Rule." *Home — Johns Hopkins All Children's Hospital*, www.hopkinsall-childrens.org/Patients-Families/Health-Library/HealthDocNew/The-5-Second-Rule. Accessed 18 Sept. 2023.

"12 Nutty Facts†about Squirrels." *Accelerator*, www.reconnectwithnature.org/news-events/big-features/12-nutty-facts-about-squirrels/. Accessed 14 Aug. 2023.

"13 Awesome Facts about Bats." *U.S. Department of the Interior*, 26 Oct. 2022, www.doi.gov/blog/13-facts-about-bats.

"44 Years Later, the Truth about the 'Science Club.'" *The New York Times*, The New York Times, 31 Dec. 1993, www.nytimes.com/1993/12/31/us/44-years-later-the-truth-about-the-science-club.html.

Andrew, Scottie. "Swear Words in Different Languages Have One Thing in Common." *CNN*, Cable News Network, 10 Dec. 2022, www.cnn.com/2022/12/10/world/swear-word-similarities-cec/index.html.

Bauza, Vanessa. "Methuselah: Still the World's Oldest Tree?" *Conservation International*, 6 June 2022, www.conservation.org/blog/methuselah-still-the-worlds-oldest-tree.

"Bed Bug Bite." *Seattle Children's Hospital*, 30 Dec. 2022, www.seattlechildrens.org/conditions/a-z/bed-bug-bite/.

Benson, Thor. "Pee Makes for Great Fertilizer. but Is It Safe?" *Popular Science*, 5 July 2022, www.popsci.com/environment/urine-fertilizer-agriculture/.

Bhatia, Aatish. "The Loudest Sound Ever Heard." *Discover Magazine*, 26 Apr. 2020, www.discovermagazine.com/environment/the-loudest-sound-ever-heard.

Bittel, Jason. "This Tiny Chameleon May Be World's Smallest Reptile." *Animals*, National Geographic, 4 May 2021, www.nationalgeographic.com/animals/article/tiny-chameleon-smallest-reptile-discovered-madagascar.

"Blue-Eyed Humans Have a Single, Common Ancestor." *ScienceDaily*, ScienceDaily, 31 Jan. 2008, www.sciencedaily.com/releases/2008/01/080130170343.htm.

"Botanists Discover New Rat-Eating Plant." *CNN*, Cable News Network, www.cnn.com/2009/WORLD/asiapcf/08/18/meat.eating.plant/. Accessed 29 Oct. 2023.

Bradford, Alina. "Facts about the Common Opossum." *LiveScience*, Purch, 21 Sept. 2016, www.livescience.com/56182-opossum-facts.html.

Bradley, Sarah. "When Do Babies Get Tears? Your Newborn May Cry without Them." *Healthline*, Healthline Media, 31 July 2020, www.healthline.com/health/baby/when-do-babies-get-tears.

Bringing Back the Woolly Mammoth and Other Extinct Creatures … Science, www.science.org/content/article/bringing-back-woolly-mammoth-and-other-extinct-creatures-may-be-impossible. Accessed 11 Sept. 2023.

"Can Rocks Grow? Trovants, the Mysterious Growing Rock." *Science ABC*, 7 Feb. 2023, www.scienceabc.com/nature/can-rocks-grow-trovants-the-mysterious-growing-rock.html.

"Can You Catch the Same Cold Twice?: Microbial Myths 3." *ASM.Org*, asm.org/Videos/2022/September/Can-You-Catch-the-Same-Cold-Twice-Microbial-My-1. Accessed 28 Aug. 2023.

Casselman, Anne. "Strange but True: The Largest Organism on Earth Is a Fungus." *Scientific American*, Scientific American, 4 Oct. 2007, www.scientificamerican.com/article/strange-but-true-largest-organism-is-fungus/.

Cheslak, Colleen. "Biography: Hedy Lamarr." *National Women's History Museum*, www.womenshistory.org/education-resources/biographies/hedy-lamarr. Accessed 11 Sept. 2023.

Christensen, Jen. "You Have a Doppelganger and Probably Share DNA with Them, New Study Suggests." *CNN*, Cable News Network, 25 Aug. 2022, www.cnn.com/2022/08/25/health/doppelganger-dna-study-wellness/index.html.

Columbia J-School Stabile Center. "One in 10 Local Covid Victims Destined for Hart Island, NYC's Potter's Field." *The City*, The City, 25 Mar. 2021, www.thecity.nyc/missing-them/2021/3/24/22349311/nyc-covid-victims-destined-for-hart-island-potters-field.

Dahlen, Hannah, and The Conversation. "What's Really Going on Behind Your Belly Button." *CNN*, Cable News Network, 3 Nov. 2017, www.cnn.com/2017/11/03/health/belly-button-partner/index.html#:~:text=Internally%20the%20veins%20and%20arteries,inside%20of%20the%20belly%20button.

"Do Ants Bury Their Dead?" *New Scientist*, 27 May 2020, www.newscientist.com/lastword/mg24632841-100-do-ants-bury-their-dead/.

"Dr. Shirley Jackson: Telecommunications Inventions." *Famous Black Inventors*, www.black-inventor.com/dr-shirley-jackson. Accessed 2 Oct. 2023.

Dunlop, Doug. "The Great Moon Hoax or Was It—The Joke's on Who?" *Smithsonian Libraries and Archives / Unbound*, 4 Sept. 2013, blog.library.si.edu/blog/2013/09/04/the-great-moon-hoax-or-was-it-the-jokes-on-who/.

Dzombak, Rebecca. "How Many Ants Are There on Earth? You're Going to Need More Zeros." *The New York Times*, 22 Sept. 2022, www.nytimes.com/2022/09/22/science/ants-census-20-quadrillion.html.

"'Frankenstein' Was Born during a Ghastly Vacation." *History.Com*, A&E Television Networks, www.history.com/news/frankenstein-true-story-mary-shelley. Accessed 28 Aug. 2023.

Frothingham, Scott. "Left Handers: Health Risks and Other Facts." *Healthline*, Healthline Media, 30 Apr. 2019, www.healthline.com/health/left-handers-and-health-risk#:~:text=About%2010%20percent%20of%20the,greater%20for%20left%20handers%2C%20too.

"The Frozen Falls." *Niagara Parks*, www.niagaraparks.com/things-to-do/frozen-falls/. Accessed 14 Aug. 2023.

Gallagher, Ashley. "Tough Teeth and Parrotfish Poop." *Tough Teeth and Parrotfish Poop*, 11 May 2023, ocean.si.edu/ocean-life/fish/tough-teeth-and-parrotfish-poop.

"Gene Roddenberry: Creator of Star Trek." *VA News*, 27 July 2022, news.va.gov/105640/gene-roddenberry-creator-of-star-trek/.

"Glass Beach." *Visit Mendocino County*, 23 Apr. 2023, www.visitmendocino.com/listing/glass-beach/.

Goldman, Jason G. "Evolution: Why Do We Have Lips?" *BBC Future*, BBC, 24 Feb. 2022, www.bbc.com/future/article/20150112-why-do-we-have-lips.

Gorvett, Zaria. "You Are Surprisingly Likely to Have a Living Doppelganger." *BBC Future*, BBC, 28 Feb. 2022, www.bbc.com/future/article/20160712-you-are-surprisingly-likely-to-have-a-living-doppelganger.

Hartzman, Marc. "Le Petomane and His Fantastic Farts at the Moulin Rouge." *Weird Historian*, 26 Mar. 2020, www.weirdhistorian.com/le-petomane/.

History of the Walter E. Fernald Development Center, www.city.waltham.ma.us/sites/g/files/vyhlif6861/f/file/file/fernald_center_history.pdf. Accessed 23 Oct. 2023.

"How Do You Clean Up 170 Million Pieces of Space Junk?" *Federation of American Scientists*, fas.org/publication/how-do-you-clean-up-170-million-pieces-of-space-junk/. Accessed 21 Aug. 2023.

"How Fast Can a Bullet Go?" *Geophysical Institute*, www.gi.alaska.edu/alaska science forum/how-fast-can-bullet-go. Accessed 14 Aug. 2023.

"How Long Does It Take Garbage to Decompose?" *Northeast Illinois Council*, storage.neic.org/event/docs/1129/how_long_does_it_take_garbage_to_decompose.pdf. Accessed 28 Aug. 2023.

"How Many Organs in the Body Could You Live Without?" *BBC Science Focus Magazine*, www.sciencefocus.com/the-human-body/how-many-organs-in-the-body-could-you-live-without. Accessed 28 Aug. 2023.

"If a Boy Scout Can Get Nuclear Materials, What's Stopping Terrorists?" *PBS*, Public Broadcasting Service, 8 Feb. 2011, www.pbs.org/newshour/science/building-a-better-breeder-reactor-1.

Impey, Chris. "3 Reasons Why the Scariest Things in the Universe Are Black Holes." *Astronomy Magazine*, 18 May 2023, www.astronomy.com/science/3-reasons-why-black-holes-are-the-scariest-things-in-the-universe/.

"It's All Smoots and Garns." *Physics World*, 11 Aug. 2023, physicsworld.com/a/its-all-smoots-and-garns/.

"Jellyfish Kill More People than Sharks Due to Their Evolved Toxin Genes." *Earth.Com*, 22 Oct. 2020, www.earth.com/news/jellyfish-kill-toxin-genes/.

"Komodo Dragon." *Smithsonian's National Zoo*, 10 Mar. 2023, nationalzoo.si.edu/animals/komodo-dragon.

Leigh-Hewitson, Nadia. "Human Hair Is Being Used to Clean Up Oil Spills." *CNN*, Cable News Network, 19 May 2022, www.cnn.com/2022/05/19/world/oil-spills-human-hair-matter-of-trust-spc-scn-intl-c2e/index.html.

"List of Continents by Population." *Continents by Population 2020—StatisticsTimes.Com*, statisticstimes.com/demographics/continents-by-population.php. Accessed 21 Aug. 2023.

Lobell, Jarrett A. "Oldest Bog Body." *Archaeology Magazine*, www.archaeology.org/issues/117-features/top10/1580-peat-bog-body-cashel-ireland. Accessed 11 Sept. 2023.

Lowe, Joe. "Birds Flying into Windows? Truths about Birds & Glass Collisions from ABC Experts." *American Bird Conservancy*, 5 Apr. 2023, abcbirds.org/blog/truth-about-birds-and-glass-collisions/.

Matt, Georg E, et al. "When Smokers Move out and Non-Smokers Move in: Residential Thirdhand Smoke Pollution and Exposure." *Tobacco Control*, U.S. National Library of Medicine, Jan. 2011, www.ncbi.nlm.nih.gov/pmc/articles/PMC3666918/.

MD Anderson Cancer Center, and Kellie Bramlet Blackburn. "What Happens When You Overeat?" *MD Anderson Cancer Center*, 25 Oct. 2019, www.mdanderson.org/publications/focused-on-health/What-happens-when-you-overeat.h23Z1592202.html.

"Meet Pando, One of the Oldest Organisms on Earth." *Earth.Com*, 27 Oct. 2020, www.earth.com/news/pando-oldest-organisms/.

Modern Mongolia: Reclaiming Genghis Khan, www.penn.museum/sites/mongolia/section2a.html. Accessed 11 Sept. 2023.

Mortensen, John. "Will the Internet Ever Run out of Space?" *Tech Evaluate*, techevaluate.com/will-the-internet-ever-run-out-of-space/. Accessed 29 Oct. 2023.

My Dinosaurs. "Top 10 Fastest Dinosaurs Ever Lived on Earth." *My Dinosaurs*, 28 May 2022, www.mydinosaurs.com/blog/top-10-fastest-dinosaurs-ever-lived-earth/.

Nall, Rachel. "Innie Belly Button vs. Outie Belly Button: Causes & More." *Healthline*, Healthline Media, 3 Aug. 2020, www.healthline.com/health/what-causes-an-innie-or-outie-belly-button#changes-later-in-life.

"National Inventors Hall of Fame Inductee Marion Donovan, Who Invented Diapers." *National Inventors Hall of Fame Inductee Marion Donovan, Who Invented Diapers*, www.invent.org/inductees/marion-donovan. Accessed 2 Oct. 2023.

"Noise Pollution Is a Major Problem, Both for Human Health and the Environment." *European Environment Agency*, 11 May 2021, www.eea.europa.eu/articles/noise-pollution-is-a-major.

Oaklander, Mandy. "Why Do We Cry? The Science of Crying." *Time*, 16 Mar. 2016, time.com/4254089/science-crying/.

"Oldest Known Rock on Earth Discovered- All Images: NSF—National Science Foundation." *NSF*, www.nsf.gov/news/news_images.jsp?cntn_id=112299&org=NSF. Accessed 24 Sept. 2023.

Onur. "What Are Tears Made of?: Cleveland Eye Clinic." *Three Types of Tears Comments*, 14 Sept. 2023, www.clevelandeyeclinic.com/2022/10/05/three-types-of-tears/.

"Overlooked No More: Bette Nesmith Graham, Who Invented Liquid Paper." *The New York Times*, 11 July 2018, www.nytimes.com/2018/07/11/obituaries/bette-nesmith-graham-overlooked.html.

"Oyster Myths & Facts (May 2023)." *In A Half Shell*, www.inahalfshell.com/oyster-myths-facts. Accessed 23 Oct. 2023.

Peterson, Elizabeth. "Walking Dead: How Wasp Overlords Control Spider Zombies." *LiveScience*, 5 Aug. 2015, www.livescience.com/51764-wasp-spider-zombies.html.

"Piltdown Man." *Natural History Museum*, www.nhm.ac.uk/our-science/departments-and-staff/library-and-archives/collections/piltdown-man.html. Accessed 18 Sept. 2023.

"The Platypus: What Nature's Weirdest Mammal Says about Our Origins", *New Scientist*, www.newscientist.com/article/mg25033332-400-the-platypus-what-natures-weirdest-mammal-says-about-our-origins/. Accessed 25 Sept. 2023.

"Rabies." *World Health Organization*, World Health Organization, www.who.int/newsroom/fact-sheets/detail/rabies#:~:text=Overview,rabies%20virus%20transmission%20to%20humans. Accessed 29 Oct. 2023.

Remple, Aidan. "What Is the Biggest Star Ever Found?" *WorldAtlas*, 1 Oct. 2022, www.worldatlas.com/space/what-is-the-biggest-star-ever-found.html.

Roland, James. "Where Does Snot Come From?" *Healthline*, Healthline Media, 29 Sept. 2018, www.healthline.com/health/where-does-snot-come-from#:~:text=Your%20nose%20and%20throat%20are,your%20nose%20and%20sinuses%20moist.

"Rutgers Researchers Debunk 'Five-Second Rule': Eating Food off the Floor Isn't Safe." *Rutgers University*, www.rutgers.edu/news/rutgers-researchers-debunk-five-second-rule-eating-food-floor-isnt-safe. Accessed 18 Sept. 2023.

"The Sailing Stones of Death Valley." *National Park Foundation*, www.nationalparks.org/connect/blog/sailing-stones-death-valley. Accessed 11 Sept. 2023.

"The Samuel George Morton Cranial Collection." *Expedition Magazine*, www.penn.museum/sites/expedition/the-samuel-george-morton-cranial-collection/. Accessed 28 Aug. 2023.

Sanderson, Katharine. "Stinky Rocks Hide Earth's Only Haven for Natural Fluorine." *Nature News*, Nature Publishing Group, 11 July 2012, www.nature.com/articles/ nature.2012.10992.

Savchuk, Alex. "Can Sound Kill You: How Much Sound Can Kill You on the Spot." *Decibel Meter App*, 14 Oct. 2021, decibelpro.app/blog/can-sound-kill-you/#:~:text=If%20we're%20talking%20about,inner%20organs%20and%20cause%20death.

Silverstein, Ken. "The Radioactive Boy Scout." *Harper's Magazine*, 18 Oct. 2012, harpers.org/archive/1998/11/the-radioactive-boy-scout/.

"Simian Crease." *Mount Sinai Health System*, www.mountsinai.org/health-library/symptoms/simian-crease. Accessed 16 Oct. 2023.

Srini Pillay, MD. "Why You Can't Get a Song out of Your Head and What to Do about It." *Harvard Health*, 4 Oct. 2017, www.health.harvard.edu/blog/why-you-cant-get-a-song-out-of-your-head-and-what-to-do-about-it-2017100412490.

Stephanie Gibeault, MSc. "Why Does My Dog Sniff Everything? Making Sense of Scents & Noses." *American Kennel Club*, 25 July 2023, www.akc.org/expert-advice/training/why-does-my-dog-sniff-everything/.

Suggitt, Connie. "56-Year-Old Freediver Holds Breath for Almost 25 Minutes Breaking Record." *Guinness World Records*, 12 May 2021, www.guinnessworldrecords.com/news/2021/5/freediver-holds-breath-for-almost-25-minutes-breaking-record-660285.

"What Happens If Someone Catches the Loch Ness Monster?" *BBC News*, BBC, 5 July 2018, www.bbc.com/news/uk-scotland-highlands-islands-44519189.

"What Happens in My Body When I Vomit?" *BBC Science Focus Magazine*, www.sciencefocus.com/the-human-body/what-happens-in-my-body-when-i-vomit. Accessed 16 Oct. 2023.

"What If the Earth Suddenly Turned Flat?" *Gizmodo*, 19 June 2017, gizmodo.com/what-if-the-earth-suddenly-turned-flat-1795819464.

"What If You Traveled in a Straight Line Forever?" *Real Clear Science*, www.realclearscience.com/2022/01/29/what_if_you_traveled_in_a_straight_line_forever_814351.html. Accessed 11 Sept. 2023.

"What Would Happen If You Fell into a Volcano?" *NBCNews.Com*, NBC Universal News Group, 27 June 2012, www.nbcnews.com/id/wbna47988287.

"Why Do I Fart So Much? The Science You Never Knew You Wanted to Know about Flatulence." *BBC Science Focus Magazine*, www.sciencefocus.com/the-human-body/fart. Accessed 11 Sept. 2023.

"Why We Don't See the Same Colors." *Psychology Today*, Sussex Publishers, www.psychologytoday.com/us/blog/the-superhuman-mind/202006/why-we-dont-see-the-same-colors. Accessed 14 Aug. 2023.

Wilcox, Christie. "How a Wasp Turns Cockroaches into Zombies." *Scientific American*, 1 May 2017, www.scientificamerican.com/article/how-a-wasp-turns-cockroaches-into-zombies1/.

"The World's Population by Eye Color." *WorldAtlas*, 31 Jan. 2023, www.worldatlas.com/society/the-world-s-population-by-eye-color.html.

Wu, Katherine J. "Elephants Can Use Scent to Distinguish More from Less." *PBS*, Public Broadcasting Service, 3 June 2019, www.pbs.org/wgbh/nova/article/elephants-smell-quantity/.

Index

Note: (ill.) indicates photos and illustrations.

Whewell, William, 93
White House, 102
Whitson, Peggy, 151
WHO (World Health Or-
ganization), 29
whorls, hair, 226
Wiccan (William Kaplan-
Altman) [comic book
character], 49
Wickman, Ivar, 101
William, Prince, 121
William II, King (William
Rufus), 34
Williams, Anna Wessels,
269
Williams, Arizona, 73
Williams, Sunita Lyn,
151, 151 (ill.)
Williams, Ted, 145
Williams Bay, Wisconsin,
15
Williamson, Ernest, 147
Williamson, George, 147
Wilson, Edward O., 242
Wimbledon, 259
Wisconsin, 15
Wish Willy. See Komodo
dragons

wolf eels, 146
Wolverine [comic book
character], 48–49
Wolverton, Bill, 271
women
first black female
engineer at NASA,
245–47
first black female
graduate of MIT,
198
first to be admitted
to MIT, 3
first to break the
sound barrier, 3–4
first to show her
navel on TV, 276
first to win a Nobel
Prize, 4–5
most spacewalk time
for, 151
Neanderthal, 6
right to vote for
black, 118
second black female
nuclear physicist,
198

Women's Army Corps,
222
Wonder Woman [comic
book character], 48
woolly mammoth, 32,
127, 127 (ill.)
World Health Organiza-
tion (WHO), 29
World War I, 181, 200
World War II, 86, 139–
40, 199, 245, 254,
263, 278
World's Fair, 122, 201
Wyman, Jane, 27
Wyoming, 274
Wyoming Dinosaur
Center and Dig Sites,
274

Y–Z

Yale University, 198
Yerkes Observatory, 15
Young Avengers [comic
book characters], 49
zombies, 187–91, 188
(ill.), 190 (ill.)